Safety and Security Engineering VIII

WITPRESS

WIT Press publishes leading books in Science and Technology.
Visit our website for new and current list of titles.
www.witpress.com

WITeLibrary

Home of the Transactions of the Wessex Institute.
Papers published in this volume are archived in the WIT eLibrary in volume 237 of
WIT Transactions on The Built Environment (ISSN 1743-3509).
The WIT eLibrary provides the international scientific community with immediate and
permanent access to individual papers presented at WIT conferences.
Visit the WIT eLibrary at www.witpress.com/elibrary.

Eighth International Conference on Safety and Security Engineering

SAFE VIII

Honorary Chairman

M. Guarascio
University of Rome, "La Sapienza", Italy

Conference Chairmen

G. Passerini
Marche Polytechnic University, Italy
Member of WIT Board of Directors

F. Garzia
University of Rome "La Sapienza", Italy

M. Lombardi
University of Rome "La Sapienza", Italy

International Scientific Advisory Committee

Sponsored by

WIT Transactions on the Built Environment
Foundation for the Basilica of Saint Francis in Assisi
International Journal of Safety and Security Engineering

Organised by

Wessex Institute, UK
University Politecnica delle Marche, Italy
University of Rome "La Sapienza", Italy

WIT Transactions

Wessex Institute
Ashurst Lodge, Ashurst
Southampton SO40 7AA, UK

Editorial Board

Safety and Security Engineering VIII

Editors

G. Passerini
Marche Polytechnic University, Italy
Member of WIT Board of Directors

F. Garzia
University of Rome "La Sapienza", Italy

M. Lombardi
University of Rome "La Sapienza", Italy

WIT PRESS Southampton, Boston

Editors:

G. Passerini
Marche Polytechnic University, Italy
Member of WIT Board of Directors

F. Garzia
University of Rome "La Sapienza", Italy

M. Lombardi
University of Rome "La Sapienza", Italy

Published by

WIT Press
Ashurst Lodge, Ashurst, Southampton, SO40 7AA, UK
Tel: 44 (0) 238 029 3223; Fax: 44 (0) 238 029 2853
E-Mail: witpress@witpress.com
http://www.witpress.com

For USA, Canada and Mexico

Computational Mechanics International Inc
25 Bridge Street, Billerica, MA 01821, USA
Tel: 978 667 5841; Fax: 978 667 7582
E-Mail: infousa@witpress.com
http://www.witpress.com

British Library Cataloguing-in-Publication Data

A Catalogue record for this book is available
from the British Library

ISBN: 978-1-78466-351-3
eISBN: 978-1-78466-352-0
ISSN: 1746-4498 (print)
ISSN: 1743-3509 (on-line)

The texts of the papers in this volume were set individually by the authors or under their supervision. Only minor corrections to the text may have been carried out by the publisher.

Preface

This Volume contains some of the papers presented at the 8th International Conference on Safety and Security Engineering held in Ancona in Italy, organised by Wessex Institute in the UK, University of Rome “La Sapienza” and University Politecnica delle Marche, Italy. The Conference was sponsored by the WIT Transactions on the Built Environment and the International Journal of Safety and Security Engineering.

This eighth meeting in the Series follows the success of the first seven meetings held in Rome (2005), Malta (2007), Rome (2009), Antwerp (2011), Rome (2013), Opatija in Croatia (2015) and Rome (2017). The purpose of the Conference is to provide a forum for the presentation and discussion of the most recent research and industrial developments in the theoretical and practical aspects of safety and security engineering.

The papers address the most recent developments in the theoretical and practical aspects of these important fields, which due to their special nature, brings together in a systematic way, many disciplines of engineering, from the traditional to the most technologically advanced. The papers cover areas such as risk analysis and assessment, system safety engineering, incident monitoring and management, information and communication security, security surveillance systems, cybercrime and health care, fire safety, disaster management, emergency response, critical infrastructure protection, counter terrorism issues, human factors and behaviour, occupational health and safety, transportation safety and security, earthquake safety, modelling and experiments, safety standards and regulations, loss prevention, and many others.

The Editors are grateful to the reviewers, as well as to all authors for their contributions.

The Editors
2019

Contents

USE OF SHERPA FOR THE PREVENTION OF HUMAN ERRORS AMONG AGRICULTURAL MACHINERY USERS

MASSIMO GUARASCIO[1], MARIO FARGNOLI[1], MARA LOMBARDI[1], DANIELE PURI[2] & FABIO GARZIA[1]
[1]Faculty of Civil and Industrial Engineering of Sapienza University of Rome, Italy
[2]Department of Technological Innovations and Safety of Plants, Products and Anthropic Settlements, National Institute for Insurance against Accidents at Work, Italy

ABSTRACT

Despite the ever stricter safety regulations and standards, the number of accidents in the agricultural sector is still relevant. The reports on occupational accidents in this sector highlight that most serious and fatal accidents are related to both the use of machinery that are not in conformity with safety protection requisites, and their misuse. The present study aims to investigate the latter aspect, focusing on the analysis of human error by means of the Systematic Human Error Reduction and Prediction Approach (SHERPA) method. In particular, we analyse the operators' behaviour and risk perception while performing agricultural activities with pedestrian controlled self-propelled machines. Results achieved show the lack of safety awareness, especially among self-employed and part-time agricultural workers, fostering the need for a more consistent effort to improve information and communication on safety issues among farmers.

Keywords: occupational safety, machinery safety, human error, SHERPA, human safety, agriculture.

1 INTRODUCTION

Safety in the agriculture sector still represent a relevant problem at a global level [1]–[3]. In fact, despite the great effort made in recent years by lawmakers in issuing stricter regulations concerning both occupational and machinery safety, the number of accidents and victims has not significantly decreased [4], [5]. The main factors causing this situation are related to the peculiar characteristics of the agricultural works, which can be summarized in the following main aspects: the frequent employees' turnover, the seasonal works' deadlines, the outdoor works in all weather conditions, the lack of highly trained workers, the use of obsolete working equipment, the frequent change of work sites, etc. [6], [7].

In particular, the reports on occupational accidents in this sector highlight that the majority of serious and fatal accidents are related to the use of machinery that are not in conformity with safety protection requisites, as well as to their misuse [8], [9]. In this study, we focused on the latter aspect, since the human behaviour plays a key role on the accident causality [10]. In fact, the identification and analysis of the hazardous events that might occur during working activities considering the human behaviour perspective can augment the quality of risk assessment activities, providing more effective results for risk reduction and prevention measures [11]–[13].

In literature, numerous studies addressing this goal can be found, mainly concerning the analysis of safety critical industries, as for example: nuclear plants, marine systems, railway and aviation industry, chemical plants, etc. [14], [15]. Conversely, in the agricultural sector, few studies have addressed these issues, mainly focusing on the ergonomics' aspects of human behaviour [16], [17]. Moreover, most studies addressing safety issues while using agricultural machinery concerned the use of four-wheel (or continuous track) tractors, while other working equipment, such as two-wheel tractors (2WTs), motor hoes or rotary tillers are still scarcely considered [18], [19].

www.witpress.com, ISSN 1743-3509 (on-line)
doi:10.2495/SAFE190011

The reasons of such a situation are due to the fact that compared with the four-wheel tractors, the accidents occurred while using the latter usually have a minor severity and caused less fatalities according to official statistics. However, their great diffusion also among non-professional users [20], as well as the fact that they do not need to be officially registered since they are not considered vehicles by the current legislation make it difficult to estimate the incidence of accidents and their effects properly. On these considerations, the purpose of the present study consists in investigating the human errors related to the use of two-wheel tractors by means of the Systematic Human Error Reduction and Prediction Approach (SHERPA) method [21]. Accordingly, in collaboration with authorities for occupational health and safety, farmers were interviewed in order to depict the context and define both the critical tasks and the related hazardous situations. Then, human errors were identified and assessed using SHERPA. The study brought to light the underestimation of potential risks by the operators, providing useful information for the implementation of safe solutions from both the management, and the engineering point of view.

2 BACKGROUND ANALYSIS

2.1 Context definition

In rural areas, agricultural pedestrian controlled self-propelled machines are highly diffused due to their multipurpose applications. In fact, starting from a common base consisting of an engine, a frame containing the gear and the power transmission, and a handle-bars command system, they can be used for driving, pulling or being equipped with different working tools depending on the different uses [19], [22]. More in detail, following the criteria provided by the EN 709:1997 standard [23], this type of equipment can be classified into the following categories. 2WTs are single-axle tractors (also called "walking tractors") characterized by a power take-off aimed at driving and/or pulling working tools (Fig. 1).

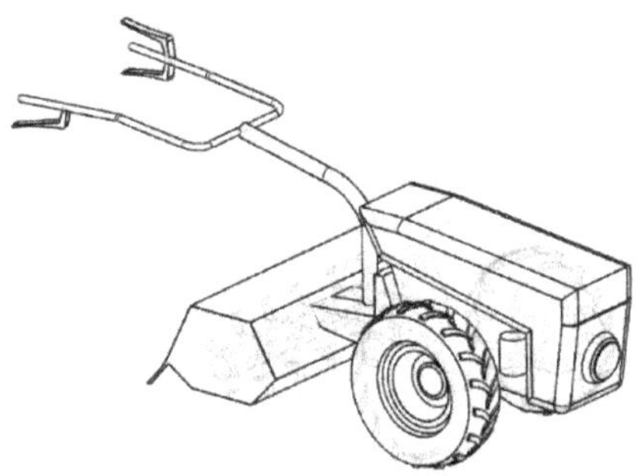

Figure 1: Scheme of a two-wheel tractor (2WT).

Instead, in motor hoes (with or without supporting wheels) the propulsion is guaranteed by the working elements directly actuated by the engine. The latter type of machinery can be also equipped with drive wheels directly actuated by the engine.

Studies concerning safety issues while using this type of machinery are not numerous and mostly addressed the problems related to vibrations. In particular, Tewari and Dewangan [24] studied the vibrations transmitted by 2WTs, proposing rubber isolators for engine mountings and handles to reduce their effects on the operators. Similarly, Chaturvedi et al. [25] reported a field analysis of vibration reduction using polyurethane and rubber isolators, as well as their combination during the different use phases of 2WTs. Guo et al. [26] analysed the handle vibrations in comparison with those at the engine cover in a motor motor-hoe. Sam and Kathirvel [27] focused their attention on hand vibrations, pointing out that also the correct use of the equipment (e.g. holding handles without straining arms, avoiding a tight grip of handles, etc.) has a significant impact on the transmission of vibrations. Accordingly, Heidary et al. [28] investigated the discomfort caused by vibrations, which may cause to operators Comfort limit and some symptoms like headache, influence on speech and lump in the throat. Besides, the comfort of operators was studied also by workloads Tiwari and Gite [29], who considered different types of discomfort while operating a 2WT leading physiological and postural.

Other studies on this type of equipment focused on their use as a transportation means especially in rural areas of developing countries, where the main concerns are related to road safety [18], [30], [31]. It has to be noted that in other countries, although is still diffused, this use of the equipment is not always allowed by safety and road traffic regulations. As for example, in Italy 2WTs and motor hoes need to undergo a specific approval in order to be considered as road mobile machinery. Otherwise, they cannot be considered as "vehicles".

2.2 Accident statistics

Although this type of equipment is usually associated to a large number of accidents among those occurred when dealing with agricultural machinery, it is very difficult to estimate their real impact on accident statistics. As a matter of fact, official statistics consider them into the category of "accidents with agricultural machinery", without specifying the type of machine/equipment involved. Besides, we also have to take into account the fact that most of the users of this type of equipment are not professional users. Moreover, we also have to consider that in most countries (e.g. in the European Union) official statistics do not include the accidents occurred to self-employed workers. Hence, it is difficult to find detailed information on this phenomenon at an official level.

Conversely, if we consider data concerning the accidents registered taking into account reports of both police departments and inspectors of the national health service, numbers are indeed larger. In Table 1 this type of data collected by the Observatory on accidents in agriculture of INAIL [32] related to the period 2013–2017 are shown.

Table 1: Accident related to the use of agricultural pedestrian controlled self-propelled machines [32].

Years	Serious injuries	Fatalities
2013	50	14
2014	34	10
2015	31	20
2016	28	11
2017	28	12
Total	171	67

Hence, on the one hand, this remarkable difference in numbers between data registered for regular employees and observatory data shows that this phenomenon and its impact on farmers is underestimated.

On the other hand, despite data provided by the observatory cannot be considered significant from the statistical point of view [32], they can provide qualitative information that can be useful to understand the characteristics of the phenomenon. More in detail, qualitative information on the number of the registered accidents classified based on the age of victims is shown in Fig. 2, while the type of accidents, i.e. the circumstances or mode in which the accident occurred in the period 2013–2017 is reported in Table 2 [32].

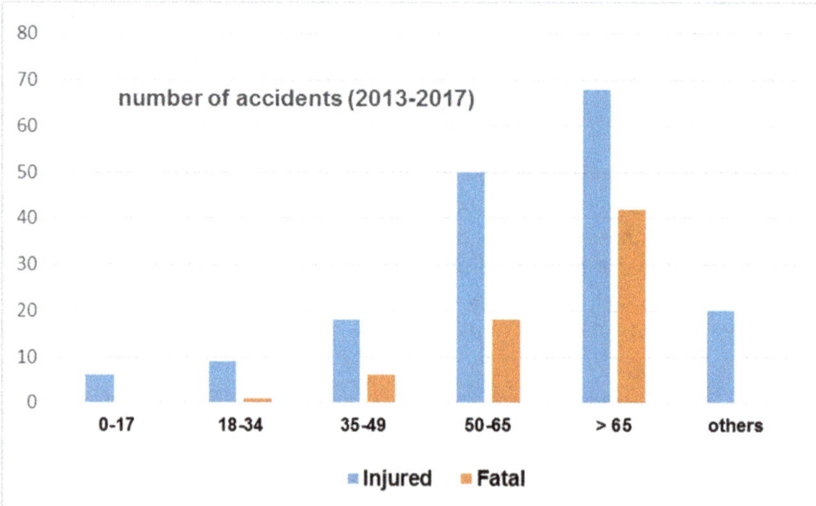

Figure 2: Number of accidents registered by the observatory of INAIL divided according to the age of victims. *(Source: Adapted from [32].)*

Table 2: Circumstances or mode of the accident in the period 2013–2017. *(Source: Adapted from [32].)*

Circumstances	Injuries	Fatal accidents
Contact/entanglement with moving parts	146	46
Crashing	8	10
Roll over	12	9
Impact with objects	1	1
Others	4	1
Total	171	67

2.3 Research issues

In such a context, despite the large number of accidents occurring to farmers needing safety and reliability analyses of human functions are needed to allow an improvement of performances and well-being [34]. most studies concerned the use of traditional tractors (i.e. four-wheel tractors), and the misperception of their safe conditions by users [35]–[39]. Other studies remarkably focused on the proper comprehension of pictograms [40] and safety signs [41] on agricultural machinery.

Instead, the analysis of the operators' behaviour and risk perception while performing agricultural activities with pedestrian controlled self-propelled machines has been scarcely investigated in literature. To reduce this gap, in this study the Systematic Human Error

Reduction and Prediction Approach (SHERPA) method [21] was used with the goal of identifying human error types and predicting their occurrence with reference to the use of pedestrian controlled self-propelled machinery.

3 RESEARCH APPROACH

3.1 The SHERPA method

More in detail, following the SHERPA method a distinction of the possible human failures can be made into the following categories [21]:

1. Action errors (A), when the human action changes the status of the system.
2. Checking errors (C), when the operator fails to verify, check the status of the system.
3. Retrieval errors (R), if the operator fails to retrieve information, e.g. from a working procedure.
4. Communication errors (C), when the operator fails to transfer information to other personnel.
5. Selection errors (S), when the operator makes the wrong choice among different alternatives.

Each one of the above categories is characterized by different error modes, as shown in Table 3.

Table 3: SHERPA error modes. *(Source: Adapted from [42].)*

Error category	Error mode	
A. Action errors	A1	Operation too long/short
	A2	Operation mistimed
	A3	Operation in wrong direction
	A4	Operation too little/much
	A5	Misalign
	A6	Right operation on wrong object
	A7	Wrong operation on right object
	A8	Operation omitted
	A9	Operation incomplete
	A10	Wrong operation on wrong object
C. Checking errors	C1	Check omitted
	C2	Check incomplete
	C3	Right check on wrong object
	C4	Wrong check on right object
	C5	Check mistimed
	C6	Wrong check on wrong object
R. Retrieval errors	R1	Information not obtained
	R2	Wrong information obtained
	R3	Information retrieval incomplete
I. Communication errors	I1	Information not communicated
	I2	Wrong information communicated
	I3	Information communication incomplete
S. Selection errors	S1	Selection omitted
	S2	Wrong selection made

In order to apply the method, an analysis of the working tasks performed is needed. As suggested by Ghasemi et al. [43], this activity is focused on bringing to light the individual's perception of task to reach goals in accordance with the working procedures and/or the instructions received. In order to perform this activity, the Hierarchical Task Analysis (HTA) method [44] was used. The output of the method consists in filling the SHERPA tables, where to each task type error modes are associated and evaluated considering the recovery possibility, the consequences, the probability of occurrence and criticality of the error (e.g. by means of a qualitative ranking such as: low, medium or high), as well as the remedial measures [21].

3.2 The case study

The case study was carried out through a series of interviews with 19 users of pedestrian controlled self-propelled machinery. More in detail, seven users of motor-hoes; three users of motor-hoes with drive wheels; and nine users of 2WTs were interviewed separately. Most of them are self-employed farmers, who uses this equipment to operate in fields where four-wheel tractors cannot be used due to space limits (e.g. in vineyards or vegetable gardens). Others are part-time agricultural workers or hobbyists. More in detail, the interviews were structured as follows:

- Definition of the main activities the task "use of the equipment" can be split in by means of a bottom up approach: i.e. the task is divided in sub-tasks; each sub-task is divided in elementary activities.
- Analysis of each activity with reference to the error classification provided in Table 3.
- Definition of error modes.

The results of the questionnaires were analysed by a group of experts in order to define the consequences of each error and the recovery opportunities of potential identified errors. Based on this an assessment of the risk level associated to each potential error was carried out using a qualitative scale (L = low; M = medium; H = high) for estimating both the probability of occurrence and the criticality of the potential errors [21].

4 RESULTS

In Fig. 3 the results of the task analysis are shown. It has to be noted that since the interviewed operators were not using exactly the same equipment, we only selected the activities that are in common among all of them, while additional operations (e.g. plowing) that can be carried out only by means of 2WTs were not taken into account. As far as the transit is concerned, the use of additional wheels was considered when the equipment is used on the road. Moreover, from the interviews it emerged that towing an agricultural trailer is quite common since 79% of interviewed operators use the equipment equipped with a trailer for reaching working areas as well as for carrying goods (e.g. wood, fruit, etc.).

For each of the above elementary activity, the error modes were identified, and the SHERPA tables were implemented accordingly: in Table 4 an excerpt of the SHERPA table concerning the use of a trailer is shown.

In total, 42 human errors were identified: 19 Action errors; six Control errors; nine Retrieval errors; two Communication errors; and six Selection errors. In Fig. 4 the status of the identified errors is shown, where the most significant ones concern the failure in completing an operation correctly.

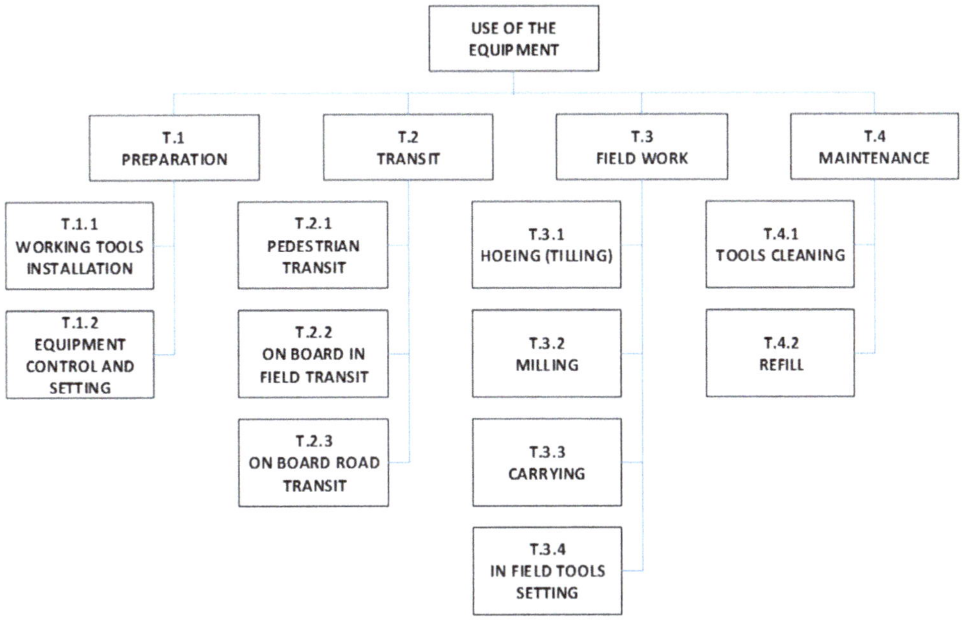

Figure 3: Hierarchical task analysis of the working activities.

Table 4: Excerpt of the SHERPA method application.

Activity	Error mode	Description	Consequence	Recovery	Probability	Criticality
T.1.1	A2	The operator installs the transit wheels without checking if the equipment is braked	Crushing, entanglement	T.2.1	L	M
	C1	The operator installs the transit wheels without checking if the equipment is cooled down	Contact with high temperature parts	T.2.1	M	L
	A6	The operator does not use the proper personal protective equipment (e.g. gloves, overalls)	Contact with cutting parts Entanglement		H	L
	A9	The operator has not fixed the wheels properly	Loss of stability Crushing Exposure to vibrations	T.2.1	L	H
	S2	The operator uses wrong transport wheels	Loss of stability Crushing Exposure to vibrations	T.2.1	L	H

Figure 4: Percentage of the identified errors. A = Action errors; C = Checking errors; R = Retrieval errors; I = Communication errors; S = Selection errors.

5 DISCUSSION OF RESULTS

The results obtained show that most of errors are action type: in particular, A2 (operation mistimed) and A7 (wrong operation on right object) are the most common in accordance with the responses of the operators. Besides, retrieval errors were also considered frequent, especially during the transportation and transit operations.

On the one hand, such an output reflects the type of operators interviewed: in fact, since they are not company employees, they do not have to follow strict working procedures. This can lead them to omit or underestimate safety issues, making it difficult to choose the right action after an error in order to bring the system to a safe condition. On the other hand, the results are in line with data provided with the INAIL's observatory [32], where most of serious/fatal accidents are related to wrong operations. This accomplishes the research clues by Cividino et al. [36], who pointed out the low perception and awareness of safety issue by the workers in the agricultural sector. Accordingly, as demonstrated also in other fields [45], [46], safety perception and climate in small working contexts need to be augmented, in order to reduce the number of accidents.

At a more general level, these results bring to light the need to improve safety information and training among farmers, considering the large number of self-employed operators and small sized companies. This is in line with the observations provided by Cavallo et al. [47], who stressed on the little attention paid to the condition of the working equipment.

Thus, the use of the SHERPA method resulted in being effective in pinpointing the safety perception and awareness of workers, bringing to light error types that can be hardly identified following a top-down analysis [48], [49]. Furthermore, such an approach can support engineers in better analysing the interactions among operators, technical systems, working environment, showing improvement options both from the technical and managerial point of view [43], [50].

Beside these positive aspects, we have to underline that the present study also presents some limitations mainly due to both the small number of interviewed operators and to the difficulties that non-expert analysts can find in applying the SHERPA method [51]. Hence, although the use of a single case-study as a research tool for exploratory investigation and to generate new understandings is recognized [52], the generalization of the research findings needs to be supported by the application of the proposed approach to further and larger analysis.

6 CONCLUDING REMARKS

Given the large number of accidents among workers in the agricultural sector especially when using working equipment, the analysis of safety perception and awareness of operators has become relevant. In such a context, the use of tools for the identification of human errors certainly represents a viable approach to augment knowledge in this research field.

This study contributes to the current research literature by proposing the application of the SHERPA method for the analysis of human errors in agricultural activities. Results achieved showed the low safety perception of self-employed and part-time workers when dealing with pedestrian controlled self-propelled machines. Such a finding can be considered exploratory and used to define new research questions and promote further and more detailed analysis.

REFERENCES

[1] Jadhav, R., Achutan, C., Haynatzki, G., Rajaram, S. & Rautiainen, R., Risk factors for agricultural injury: A systematic review and meta-analysis. *Journal of Agromedicine*, **20**(4), pp. 434–449, 2015.

[2] Zheng, L. et al., Nonfatal work-related injuries among agricultural machinery operators in northern China: A cross-sectional study. *Injury*, **45**(3), pp. 599–604, 2014.

[3] Houshyar, E. & Houshyar, M., Tractor safety and related injuries in Iranian farms. *Safety Science*, **103**, pp. 88–93, 2018.

[4] Cecchini, M. et al., Survey on the status of enforcement of European directives on health and safety at work in some Italian farms. *Journal of Food, Agriculture and Environment,* **11**(3–4), pp. 595–600, 2013.

[5] Lombardi, M. & Fargnoli, M., Prioritization of hazards by means of a QFD-based procedure. *International Journal of Safety and Security Engineering*, **8**(2), pp. 342–353, 2018.

[6] Moradhaseli, S., Farhadian, H., Abbasi, E. & Ghofranipour, F., Factors affecting the incidence of occupational accidents among farmers. *Health Education and Health Promotion*, **5**(1), pp. 39–56, 2017.

[7] Fargnoli, M., Vita, L., Gattamelata, D., Laurendi, V. & Tronci, M., A reverse engineering approach to enhance machinery design for safety. *Proceedings of DESIGN 2012,* Dubrovnik, Croatia, pp. 627–636, 2012.

[8] Mishra, D. & Satapathy, S., A framework designed for macro-ergonomical analysis of Indian farmers: assessment and analysis of occupational injuries of agricultural farmers of South Odisha in India. *Advanced Macroergonomics and Sociotechnical Approaches for Optimal Organizational Performance,* IGI Global, pp. 162–183, 2019.

[9] Robert, K., Elisabeth, Q. & Josef, B., Analysis of occupational accidents with agricultural machinery in the period 2008–2010 in Austria. *Safety Science*, **72**, pp. 319–328, 2015.

[10] Irwin, A. & Poots, J., The human factor in agriculture: An interview study to identify farmers' non-technical skills. *Safety Science*, **74**, pp. 114–121, 2015.

[11] Schwab, C.V., Mosher, G.A. & Ryan, S.J., Agricultural worker injury comparative risk assessment methodology: Assessing corn and biofuel switchgrass production systems. *Journal of Agricultural Safety and Health*, **23**(3), pp. 219–235, 2017.
[12] Narasimhan, G., Crowe, T.G., Peng, Y., Hagel, L., Dosman, J. & Pickett, W., A task-based analysis of machinery entanglement injuries among Western Canadian farmers. *Journal of Agromedicine*, **16**(4), pp. 261–270, 2011.
[13] Llory, M., Human-and work-centred safety: Keys to a new conception of management. *Ergonomics*, **40**, pp. 1148–1158, 1997.
[14] Alvarenga, M.A.B., Melo, P.F. & Fonseca, R.A., A critical review of methods and models for evaluating organizational factors in human reliability analysis. *Progress in Nuclear Energy*, **75**, pp. 25–41, 2014.
[15] Borghini, F., Garzia, F., Lombardi, M., Mete, M., Perruzza, R. & Tartaglia, R., Human factor analysis inside a peculiar job environment at the Gran Sasso mountain underground laboratory of Italian National Institute for Nuclear Physics. *International Journal of Safety and Security Engineering*, **8**(3), pp. 390–405, 2018.
[16] Tiecheng, Z., The study on the ergonomics of hand tractor based on digital human model. *Proceedings of the Sixth International Conference on Measuring Technology and Mechatronics Automation*, IEEE, pp. 778–781, 2014.
[17] Patel, T., Sanjog, J. & Karmakar, S., Ergonomics perspective in agricultural research: A user-centred approach using CAD and digital human modeling (DHM) technologies. *Journal of The Institution of Engineers (India): Series A*, **97**(3), pp. 333–342, 2016.
[18] Ericson, M., Two-wheel tractors: Road safety issues in Laos and Cambodia. *Safety Science*, **48**(5), pp. 537–543, 2010.
[19] Tiwari, P.S. & Gite, L.P., Physiological responses during operation of a rotary power tiller. *Biosystems Engineering*, **82**(2), pp. 161–168, 2002.
[20] Rathnayaka, N., Amadoru, I., Gunathilaka, H. & Chamara, A., Performance evaluation of a two wheel tractor (Sonik mini cultivator). *Proceedings of 17th Agricultural Research Symposium*, pp. 163–167, 2018.
[21] Harris, D., Stanton, N.A., Marshall, A., Young, M.S., Demagalski, J. & Salmon, P., Using SHERPA to predict design-induced error on the flight deck. *Aerospace Science and Technology*, **9**(6), pp. 525–532, 2005.
[22] Rasool, S. & Raheman, H., Improving the tractive performance of walking tractors using rubber tracks. *Biosystems Engineering*, **167**, pp. 51–62, 2018.
[23] EN 709:1997+A4:2009, Agricultural and forestry machinery. Pedestrian controlled tractors with mounted rotary cultivators, motor hoes, motor hoes with drive wheel(s). Safety. European Committee for Standardization: Geneva, 2009.
[24] Tewari, V.K. & Dewangan, K.N., Effect of vibration isolators in reduction of work stress during field operation of hand tractor. *Biosystems Engineering*, **103**(2), pp. 146–158, 2009.
[25] Chaturvedi, V., Kumar, A. & Singh, J.K., Power tiller: Vibration magnitudes and intervention development for vibration reduction. *Applied Ergonomics*, **43**(5), pp. 891–901, 2016.
[26] Guo, L., Jian, C., Hangjia, X. & Shimeng, W., Vibration test and analysis of mini-tiller. *International Journal of Agricultural and Biological Engineering*, **9**(3), pp. 97–103.
[27] Sam, B. & Kathirvel, K., Vibration characteristics of walking and riding type power tillers. *Biosystems Engineering*, **95**(4), pp. 517–528, 2006.

[28] Heidary, B., Hassan-Beygi, S.R. & Ghobadian, B., Ergonomic characteristics and operator body fatigue against two-wheel tractor vibration. *International Journal of Agriculture and Crop Sciences*, **5**(4), pp. 370–376, 2013.

[29] Tiwari, P.S. & Gite, L.P., Evaluation of work-rest schedules during operation of a rotary power tiller. *International Journal of Industrial Ergonomics*, **36**(3), pp. 203–210, 2006.

[30] Kumar, G.P. & Dewangan, K.N., Agricultural accidents in north eastern region of India. *Safety Science*, **47**(2), pp. 199–205, 2009.

[31] Kim, B., Yum, S., Kim, Y.Y., Yun, N., Shin, S.Y. & You, S., An analysis of factors relating to agricultural machinery farm-work accidents using logistic regression. *Journal of Biosystems Engineering*, **39**(3), pp. 151–157, 2014.

[32] National Institute for Insurance against Accidents at Work (INAIL), Osservatorio nazionale sugli incidenti in agricoltura (National observatory on occupational accidents in agriculture), 2018.

[33] Lombardi, M. & Rossi, G., Cluster analysis of fatal accidents series in the Infor.mo database: Analysis, evidence and research perspectives. *International Journal of Safety and Security Engineering*, **3**(4), pp. 317–331, 2013.

[34] Caffaro, F., Cremasco, M.M., Giustetto, A., Vigoroso, L., Paletto, G. & Cavallo, E., Ergonomics in agriculture: critical postures, gestures, and perceived effort in handling foldable roll-over protective structures (ROPS) fitted on tractors. *Proceedings of the Congress of the International Ergonomics Association*. Springer: Cham, pp. 194–202, 2018.

[35] Houshyar, E. & Houshyar, M., Tractor safety and related injuries in Iranian farms. *Safety Science*, **103**, pp. 88–93, 2018.

[36] Cividino, S.R.S., Pergher, G., Zucchiatti, N. & Gubiani, R., Agricultural health and safety survey in Friuli Venezia Giulia. *Agriculture*, **8**(1), p. 9, 2018.

[37] Fargnoli, M., Lombardi, M., Haber, N. & Puri, D., The impact of human error in the use of agricultural tractors: A case study research in vineyard cultivation in Italy. *Agriculture*, **8**(6), p. 82, 2018.

[38] Hollnagel, E., Reliability analysis and operator modelling. *Reliability Engineering and System Safety*, **52**(3), pp. 327–337, 1996.

[39] Irwin, A., Caruso, L. & Tone, I., Thinking ahead of the tractor: Driver safety and situation awareness. *Journal of Agromedicine*, pp. 1–10, 2019.

[40] Caffaro, F., Schmidt, S., Murphy, D.J. & Cavallo, E., Comprehension rates of safety pictorials affixed to agricultural machinery among Pennsylvania rural population. *Safety Science*, **103**, pp. 162–171, 2018.

[41] Vigoroso, L., Caffaro, F. & Cavallo, E., Warning against critical slopes in agriculture: comprehension of targeted safety signs in a group of machinery operators in Italy. *International Journal of Environmental Research and Public Health*, **16**(4), p. 611, 2019.

[42] Stanton, N.A. & Baber, C., Error by design: Methods for predicting device usability. *Design Studies*, **23**(4), pp. 363–384, 2002.

[43] Ghasemi, M., Nasleseraji, J., Hoseinabadi, S. & Zare, M., Application of SHERPA to identify and prevent human errors in control units of petrochemical industry. *International Journal of Occupational Safety and Ergonomics*, **19**(2), pp. 203–209, 2013.

[44] Stanton, N.A., Hierarchical task analysis: Developments, applications, and extensions. *Applied Ergonomics*, **37**, pp. 55–79, 2006.

[45] Khosravi, Y., Asilian-Mahabadi, H., Hajizadeh, E., Hassanzadeh-Rangi, N., Bastani, H. & Behzadan, A.H., Factors influencing unsafe behaviors and accidents on construction sites: A review. *International Journal of Occupational Safety and Ergonomics*, **20**(1), pp. 111–125, 2014.

[46] Fargnoli, M. & Lombardi, M., Preliminary Human Safety Assessment (PHSA) for the improvement of the behavioral aspects of safety climate in the construction industry. *Buildings*, **9**(3), p. 69, 2019.

[47] Cavallo, E., Ferrari, E., Bollani, L. & Coccia, M., Attitudes and behaviour of adopters of technological innovations in agricultural tractors: A case study in Italian agricultural system. *Agricultural Systems*, **130**, pp. 44–54, 2014.

[48] Hughes, C.M., Baber, C., Bienkiewicz, M., Worthington, A., Hazell, A. & Hermsdörfer, J., The application of SHERPA (Systematic Human Error Reduction and Prediction Approach) in the development of compensatory cognitive rehabilitation strategies for stroke patients with left and right brain damage. *Ergonomics*, **58**(1), pp. 75–95, 2015.

[49] Fargnoli, M., Lombardi, M., Haber, N. & Guadagno, F., Hazard function deployment: A QFD-based tool for the assessment of working tasks: A practical study in the construction industry. *International Journal of Occupational Safety and Ergonomics*, pp. 1–22, 2018.

[50] Di Pasquale, V., Miranda, S., Iannone, R. & Riemma, S., A simulator for human error probability analysis (SHERPA). *Reliability Engineering and System Safety*, **139**, pp. 17–32, 2015.

[51] Lane, R., Stanton, N.A. & Harrison, D., Applying hierarchical task analysis to medication administration errors. *Applied Ergonomics*, **37**(5), pp. 669–679, 2006.

[52] Yin, R.K., *Case Study Research: Design and Methods*, 4th ed., Sage: Thousand Oaks, CA, 2008.

EVOLUTION OF COGNITIVE DEMAND IN THE HUMAN–MACHINE INTERACTION INTEGRATED WITH INDUSTRY 4.0 TECHNOLOGIES

MARIANNA MADONNA[1], LUIGI MONICA[2], SARA ANASTASI[2] & MARIO DI NARDO[3]
[1]INAIL UOT CVR, Italy
[2]INAIL DIT, Italy
[3]Department of Materials Engineering and Operations Management, Faculty of Engineering, University of Naples "Federico II", Italy

ABSTRACT

In the highly technological and innovative scenario of Industry 4.0, characterized by a series of enabling technologies and a strong interconnection of resources, it is necessary to take into account the impact that the introduction of increasingly sophisticated sensors and collaborative machines on the safety aspects. In addition to the introduction of so-called "smart" technologies, the context of the use of new technologies and the tasks of front-line operators has also changed. The worker increasingly assumes the role of supervisor and when some types of work require particular human skills, there is a real "collaboration" between man and machine. In the new factories, the interaction with "smart machines" on one hand simplifies the operations of the worker making them less complex and less susceptible to errors and on the other hand increases the information and communication of these systems and leads to a complexity that requires new man–machine interface modes. The support of sensors and new technologies allows the detection of a series of data necessary to make the most reliable predictions on the state of health of the equipment so that it is possible to plan target interventions. This implies that the cognitive interaction effort of the machine operator moves from the skill level to the knowledge level because the human is required to manage a huge amount of data (big data) that must be acquired, analysed and interpreted. This paper, starting from consolidated human reliability methodologies in the literature, which allows for evaluating human error in different work fields, aims to highlight how human performance improves even if it implies an increase in cognitive demand due to the use of new smart technologies.

Keywords: Industry 4.0, smart technologies, human–machine interaction, human factors and behaviour.

1 INTRODUCTION

The new connectivity and the interaction technologies among the elements of the socio-technical system, which characterized Industry 4.0, modify the traditional relationship man–machine-environment. The Internet of Things (IoT) allows the organizations to gather vast amount of data from smart devices equipped on processes, machines and products so that they allow monitoring their status and taking actions in real-time. The enabling technologies, such as virtual and augmented reality, collaborative robots, exoskeletons, support and aid human work. However, the advanced technologies reduce operator's physical workload, also allowing humans to perform their job more safely, but they require new demands of operators' skills and competencies. In fact, the control of intelligent machines requires a new cognitive load as the lower skilled human jobs were replaced by advanced technology while many high precision and complex processes are still managed by humans or by robots collaboration thanks to advanced human–machine interface (HMI) [1].

In this new scenario, the key factor for 4.0 competitiveness of companies will be the presence of highly qualified personnel, able to guarantee not only the implementation of new technologies but also the correct supervision of plant processes by the latter.

WIT Transactions on The Built Environment, Vol 189, © 2019 WIT Press
www.witpress.com, ISSN 1743-3509 (on-line)
doi:10.2495/SAFE190021

The paper aims, in the first section, to describe the new role of the front-line operator in the smart factories and in the second section to define a conceptual cognitive framework of operator 4.0.

2 EVOLUTION OF THE ROLE OF FRONT-LINE OPERATOR OVER TIME

According to Romero et al. [2], the role of the operator in production has changed following the industrial revolutions that have taken place over the time, as shown in Fig. 1.

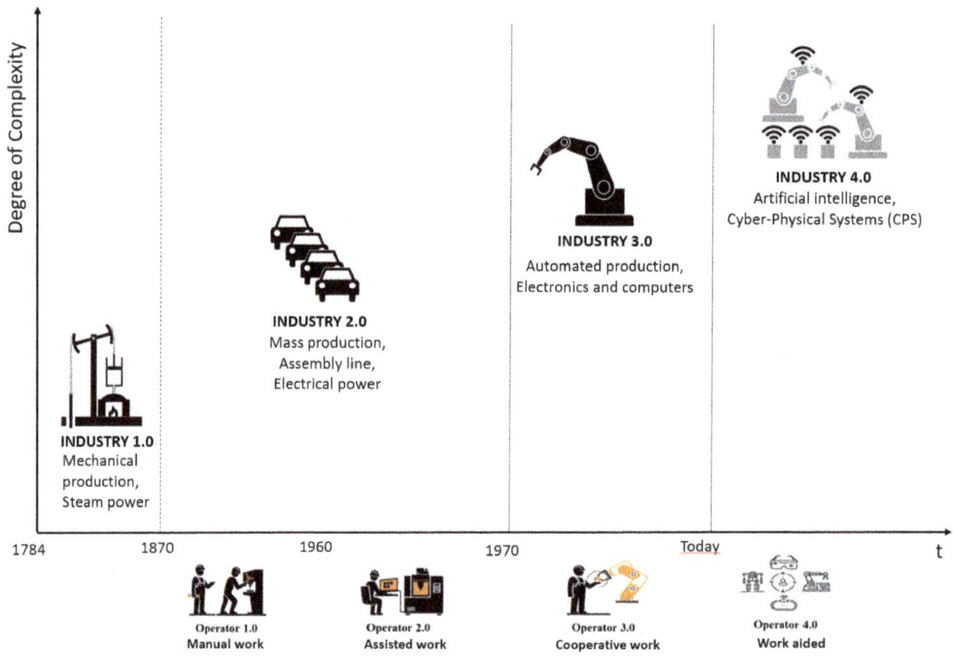

Figure 1: Operator evolution according to industrial revolutions. *(Adapted from Romero et al. [2].)*

Each industrial revolution is characterized by an increasing degree of complexity of the technologies involved and by a different role of the front-line operator who works with them. The *Operator 1.0* generation is defined in [2] as "humans conducting manual and dextrous work with the support from mechanical tools and manually operated machine tools". The *Operator 2.0* performs "assisted work" with the support of computer tools (e.g. Computer Numerical Control) while the *Operator 3.0* performs "cooperative work" with robots and other machine and computer tools, also known as human robot collaboration. The *Operator 4.0* is "the operator of the future", a smart and skilled operator whose physical, sensory and cognitive capabilities are enhanced using advanced enabling technologies.

2.1 The new role of operator in smart factory

As the smart factory is characterized by an increasingly need for flexibility and adaptability of production systems, an increased cognitive workload of the operator 4.0 is required (e.g. diagnosis, situation awareness, decision-making, planning, etc.).

According to the vision of the operator 4.0 in Romero et al. [3], that depicts how the enabling technologies in the Industry 4.0 enhance the operator in carry out the task, eight types of operators are identified; they interact physically and cognitively with the technology:

1. *Super-Strength Operator* who uses exoskeleton;
2. *Collaborative Operator* who uses a collaborative robot;
3. *Virtual Operator* who uses virtual reality;
4. *Augmented Operator* who uses augmented reality;
5. *Smarter Operator* who uses intelligent personal assistant;
6. *Social Operator* who uses social networks;
7. *Analytical Operator* who uses Big Data analytics;
8. *Healthy Operator* who uses a wearable tracker [4].

In this paper, focused on human–machine interface, only cognitive interactions are considered:

Smarter Operator: an operator helped by an intelligent personal assistant (software agent or artificial intelligence) in performing his tasks handsfree by means of voice interaction technology. This interaction has positive repercussions on the safety of the operator as it reduces the error probability especially in searching and retrieving from a digital library through voice commands. Many advantages are also in maintenance activities when the instructions are read to the operator while he performs tasks.

Virtual Operator: an operator aided by an advanced simulation of realistic environment so that some activities are carried out in safer manner. For example, virtual reality can help operator to perform tasks, by reducing error probability as it improves operator training on new assembly and maintenance procedures.

Augmented Operator: an operator whose field of view of the real factory environment is enriched, by means of smartphones, tablet, etc., with digital information. This technology offers many advantages in terms of reducing human errors as it supports operator in carrying out his task by becoming digital assistance and improving decision-making when feedbacks are displayed in real-time.

Analytic Operator: this operator uses big data analytics, the process of collecting, organizing and analyzing amount of data in order to gather useful information to predict events. The increasing available data by means of cheap sensors and Internet of Things allow the operator to receive in real time warnings about machine fault and to make predictions more accurately [4].

3 THE EVOLUTION OF COGNITIVE DEMAND

The human reliability analysis (HRA) methods have been developed to quantify human error probability in performing an assigned task, so they needed to formalize a model of human behaviour that describes the cognitive process of human beings and its link with human performance [5].

The aim of this paper is to show how human performance improves even if it implies an increase in cognitive demand due to the use of new smart technologies, therefore there is the need to take as reference a cognitive model.

Therefore the model considered is that underlying the Cognitive Reliability and Error Analysis Method (CREAM), one of the second generation of HRA techniques [6].

Earlier versions of the CREAM made use of a simplified model of cognition called SMoC, which just means Simple Model of Cognition (Hollnagel and Cacciabue [7]).

The purpose of SMoC was to describe the human cognition based on four basic cognitive functions: observation, interpretation, planning and execution. The main feature of the SMoC is the cyclical nature of human cognition, as it implied atypical path from observation over interpretation and planning to execution. It can generate any sequential model, including the well-known *step-ladder* cognitive model (Rasmussen [8]), which is structured on three different pathways of increasing complexity that require increasing levels of attention and cognitive resources.

According to mental model described in CREAM, cognition should not be described as a sequence of steps, but as a controlled use of the available competence (skills, procedures, and knowledge) and resources. So that, each typical cognitive activity can be described in terms of combination of the four cognitive functions it requires.

The cognitive profile is based on a table of the cognitive functions associated with each of the cognitive activity as shown below in Table 1, based on a Simple Model of Cognition. The following list of cognitive activities derived from several source based on assumption that they are exhaustive in pragmatic sense but from analytical point of view it is impossible to prove that they are complete, consistent, or even correct.

Table 1: A generic cognitive activity by cognitive demand matrix. *(Adapted from [6].)*

Activity type	**Cognitive function**			
	Observation	**Interpretation**	**Planning**	**Execution**
Co-ordinate			X	X
Communicate				X
Compare		X		
Diagnose		X	X	
Evaluate		X	X	
Execute				
Identify		X		
Maintain			X	X
Monitor	X	X		
Observe	X			
Plan			X	
Record		X		X
Regulate	X			X
Scan	X			
Verify	X	X		

As the first CREAM step, we want to build a *profile of the cognitive demands* of one of type of human-enabling technology interaction as characterized by Romero et al. [3]. In particular, in order to put into practice the construction of the cognitive profile, an operator intervention by means of augmented reality in a maintenance activity [9], is considered.

3.1 The cognitive demand in human-enabling technology interaction in maintenance activity: an example

An operator in a shop floor monitors the health status of the machines and the process parameters via display. At some point, an alarm is displayed on his control station. The alarm refers to a fault on a machine that forms paper envelopes. Since the machine is equipped with sensors, they transmit signals on its status to operator's workstation. On the display, the machine scheme is shown in its entirety and the fault is precisely located. The area of the machine affected by the fault that led to the machine down, is highlighted in red colour, precisely the area between the unfolding of the paper roll and the entrance of it into the rollers. The operator is trained to intervene with the support of augmented reality. The display of his workstation is that of a portable tablet. He removes it from the fixed base and moves himself in front of the machine where the fault occurred. The operator selects the augmented reality mode on the tablet. In this way, as augmented reality has been applied to all the machines in the shop-floor, when the operator photographs the machine, a library opens up display where it is possible to select an interactive use and maintenance manual. Furthermore, by placing the tablet's camera towards the machine, a series of additional information are displayed such as danger and warning signs relating to the dangerous areas of the machine and the PPE (personal protective equipment) necessary to operate on it. The operator puts the suggested PPE and takes the labelled toolbox to operate on that machine.

The operator selects from the library, which has a set of on-line tools [10], the use and maintenance manual and in particular the procedure for restoring the functionality in that area of the machine affected by the paper jam.

The selected procedure is explained step by step and simultaneously shown on the display with the appropriate danger warnings. The operator adheres to the procedure in the execution and completes it successfully and safely.

In Table 2, the maintenance activity, described above, has been divided into elementary tasks. The relative cognitive activity was associated to each task and therefore, through Table 1, the cognitive functions involved are identified.

Table 2: Cognitive demands table for maintenance activity aided with augmented reality.

Task step or activity	Cognitive activity	Obs	Int	Plan	Exe
Keep track of machine state over the time	Monitor	x	x		
Establish the identity of machine state	Identify		x		
Assess the actual situation based on available information	Evaluate		x	x	
Organise a set of actions by which the repair will be successfully achieved	Plan			x	
Perform specific actions according to step by step procedures displayed on tablet	Execute				x
Check the feedback from prior operations	Verify	x	x		

As result, the cognitive demand in this maintenance activity aided with augmented reality, implies all four cognitive functions, with prevalence of the interpretation function, followed by observation and planning and finally by execution. Therefore, it is possible to deduce that the use of advanced technologies of Industry 4.0 requires a considerable cognitive commitment. However, it is worth highlighting how the use of augmented reality offers the operator a real-time training with step-by-step procedures such as to minimize errors due to the incorrect or omitted execution of one of the elementary tasks.

4 A CONCEPTUAL FRAMEWORK OF THE OPERATOR 4.0 COGNITIVE MODEL

In Fig. 2, the cognitive model of operator in smart factory is schematized in a conceptual framework. A huge quantity of data is collected from physical world by means of sensors on equipment, processes and products. By means of IoT, data are transmitted and displayed on the operator interface technologies. In this way, the operator become a data receiver of the Cyber–Physical System (CPS), the virtual twin of physical system, keeping the process parameters and the health status of the machines under control. The data collected and displayed on the human technology interface are processed, first implying the cognitive functions of observation and interpretation and after requiring knowledge-based activities that involve the planning and the execution of the consequent actions. The new human–machine interaction implies all four cognitive functions and the advanced technologies enhance operator to perform tasks more reliably. In fact, the data related to process status are more reliable (so the observation and interpretation processes are less susceptible to human error) and the augmented and virtual reality assist the operator in the process of decision making (planning) and in acting (execution) in a safer way thanks to procedures dictated step by step.

5 CONCLUSION

The new cognitive demand of front-line operator is due to the increasing complexity of the information displayed, the need to interpret the content of such information and, if necessary,

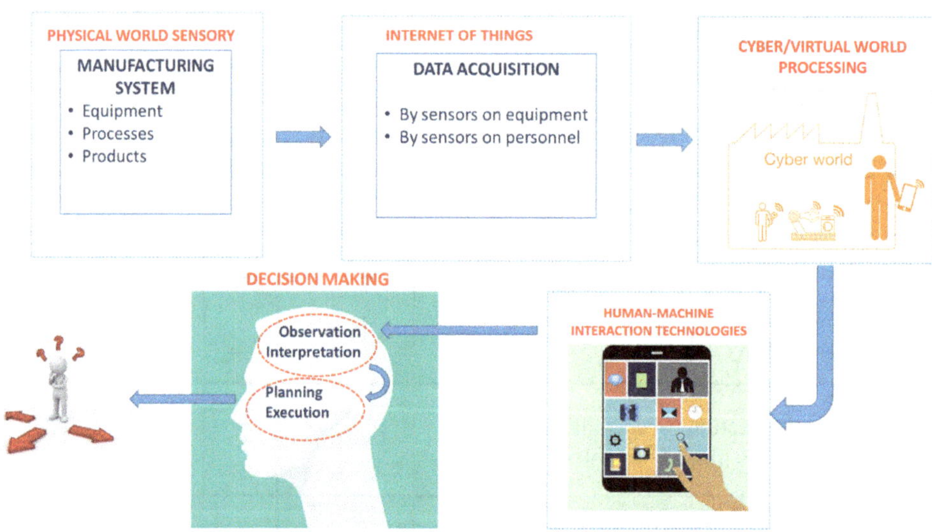

Figure 2: A conceptual cognitive framework of operator in smart factory.

to promptly act. For this reason, the HMI plays an important role, as it has to guide and teach the user in a virtual or augmented environment. As worker's competence profile changes, teaching and training are necessary in updating the skills required.

In the future, it would be interesting to compare the probability of human error of an operator who works in a traditional way with that of an operator using enabling Industry 4.0 technologies. The assessment of the human probability error (HPE) could be performed using one of the human reliability techniques available in literature. In accordance with the analysis conducted in this paper, the CREAM is well suited for this purpose.

Furthermore, another aspect to consider is that the virtual and the augmented reality could be useful tools in training, as they allow the operator to act in dynamic and unforeseen situation safely also by means of on-line step-by-step procedures [11].

In this highly innovative scenario, the human resource requires more competences and skills to manage new technologies. Therefore, the challenge that companies will have to face will be *knowledge management* that becomes another strategic innovation factor.

REFERENCES

[1] Peruzzini, M., Grandi, F. & Pellicciari, M., Exploring the potential of Operator 4.0 interface and monitoring. *Computer and Industrial Engineering*, 2018.

[2] Romero, D., Bernus, P., Noran, O., Stahre, J. & Fast-Berlung, Å. The Operator 4.0: Human cyber–physical systems and adaptive automation towards human–automation symbiosis work system. *IFIP International Conference on Advances in Production Management Systems*, pp. 677–686, 2016.

[3] Romero, D. et al., Towards an Operator 4.0 typology: A human-centric perspective on the fourth industrial revolution technologies. *International Conference on Computers and Industrial Engineering (CIE46)*, pp. 1–11, 2016.

[4] Zolotovà, I., Papcun, P., Ajàti, E., Miskuf, M. & Mocnej, J., Smart and cognitive solutions for Operator 4.0: Laboratory H-CPPS case study. *Computer and Industrial Engineering*, 2018.

[5] Di Nardo, M., Gallo, M., Madonna, M. & Santillo, L.C. A conceptual model of human behaviour in socio-technical systems. *Communications in Computer and Information Science*, **532**, pp. 598–609, 2015.

[6] Hollnagel, E., *Cognitive Reliability and Error Analysis Method: CREAM*, Elsevier, 1998.

[7] Hollnagel, E. & Cacciabue, C., Modeling cognition and erroneous actions in system simulation contexts. *The 3rd European Meeting on Cognitive Science Approaches to Process Control*, 1991.

[8] Rasmussen, J., *Information Processing and Human–Machine Interaction: An Approach to Cognitive Engineering*, Wiley, 1986.

[9] Bottani, E. & Vignali, G., Augmented reality technology in the manufacturing industry: A review of the last decade. *IISE Transactions*, **51**(3), pp. 284–310, 2019.

[10] Lotti, G., Villani, V., Battilani, N. & Fantuzzi, C., Towards an integrated approach for supporting the workers in Industry 4.0. *Proceedings of IEEE Industrial Cyber–Physical Systems, ICPS*, 2018.

[11] Villani, V., Sabattini, L., Czerniak, J.N., Mertens, A., Vogel-Heuser, B. & Fantuzzi, C., Towards modern inclusive factories: A methodology for the development of smart adaptive human–machine interfaces. *Proceedings IEEE International Conference on Emerging Technologies and Factory Automation, ETFA*, 2017.

ANALYSIS OF THE ASSURANCE OF WORK EXPERIENCE FOR UNIVERSITY GRADUATES IN THE CONTEXT OF INDUSTRY 4.0

VENDULA LACIOK[1], ALES BERNATIK[1], LENKA SVOBODOVA[2],
IVETA MLEZIVOVA[2], VERONIKA HEJDOVA[1] & PAVEL SUSKA[1]
[1]Faculty of Safety Engineering, VSB Technical University of Ostrava, Czech Republic
[2]Occupational Safety Research Institute, Czech Republic

ABSTRACT

Since the first industrial revolution, which brought us great inventions such as steam engines and the mechanical loom, society has entered new phases of revolution. Artificial intelligence (AI) and 3D printing are being introduced into the manufacturing business and the economy, while industry is facing new challenges in the fields of cyber–physical systems and information technology (IT). The overall concept of Industry 4.0 will impact not only industry, but also the areas of safety, the education system, the legal framework, science and research, and the social system. It is already necessary to start preparing the current generation for these new challenges and to develop the potential of mankind. This document is a result of the research project TL01000470 "Potential Impacts of Industry 4.0 on Operators in 3.0 Jobs and Tertiary Education in Accordance with Safety Engineering". This research can be considered the first step in an analysis to ensure that graduates gain significant practical experience in the context of Industry 4.0. This research project focuses on analysing the current situation on the labour market from the perspective of employers and their requirements of students and university graduates in the Czech Republic. The analysis focuses on knowledge of Industry 4.0 among students at the Faculty of Safety Engineering at VSB – Technical University of Ostrava. The state of preparedness for the world of work among graduates is inadequate from the practical viewpoint and it is essential to respond to the fourth industrial revolution by transforming the university syllabus.
Keywords: Industry 4.0, safety, student.

1 INTRODUCTION

The fourth industrial revolution (Industry 4.0) is being reflected progressively in every sector of national economies. Small, medium-sized and large concerns are beginning to apply the basic tools of Industry 4.0 and are digitalising process data, while other companies are introducing collaborative robots for work with man.

The approach to Industry 4.0 differs not merely in individual production sectors, but also beyond the borders of the Czech Republic. States take advantage of different means, finances and working groups, while there is also a functioning system of co-operation and projects serving as sources of inspiration and motivation for a further shift forwards in automation, digitalisation and the world of robots. Countries represent and define their progress and goals in Industry 4.0 with the help of initiatives and national programmes.

The World Economic Forum is devoting attention to Industry 4.0 on a broad scale. Since 1979 it has been issuing an annual report on the competitiveness of countries (the Global Competitiveness Report or GCR). This GCR considers and compares the performance economy in the areas of investments and information and communication technology on the basis of an index comprised of 12 predictors. It considers the following areas: institutions, infrastructure, the macroeconomic environment, education, the market in goods, the labour market, the financial market, technological readiness, the size of the market, business sophistication and innovation. The 2016 report is devoted to the preparedness of individual states for the challenge represented by Industry 4.0 [1]. Table 1 shows the ranking of

www.witpress.com, ISSN 1743-3509 (on-line)
doi:10.2495/SAFE190031

Table 1: The ranking of states in the Global Competitiveness Report [2], [3].

Name of country	Placing in ranking (index value)	
	2016–2017	2017–2018
Switzerland	1 (5.76)	1 (5.81)
Singapore	2 (5.68)	2 (5.72)
United States of America	3 (5.61)	3 (5.70)
Netherlands	5 (5.50)	4 (5.57)
Germany	4 (5.53)	5 (5.57)
Sweden	9 (5.43)	6 (5.53)
Great Britain	10 (5.43)	7 (5.49)
Japan	6 (5.47)	8 (5.48)
Finland	8 (5.45)	10 (5.44)
Belgium	19 (5.20)	17 (5.25)
Luxemburg	20 (5.20)	20 (5.20)
France	22 (5.13)	21 (5.20)
Australia	21 (5.15)	22 (5.19)
Ireland	24 (5.11)	23 (5.18)
Israel	27 (4.98)	24 (5.18)
Republic of Korea	26 (4.99)	26 (5.03)
China	28 (4.89)	28 (4.95)
Czech Republic	31 (4.69)	31 (4.72)
Russia	45 (4.44)	43 (4.51)
Slovak Republic	67 (4.22)	65 (4.28)
Ukraine	79 (4.04)	85 (4.00)

individual states in two time periods between 2016 and 2018. Switzerland, Singapore and the United States of America hold stable leading positions in the rankings. The Czech Republic in 31st place, with an index value of 4.72, is 1.09 index value points behind leading Switzerland.

If we compare the given states, Switzerland in first position and the Czech Republic, in the years 2017–2018 we can see in the following images the areas in which Switzerland is ahead (see Fig. 1). The Czech Republic can compete in the areas of health and education. Its weakest position, in contrast, is in the area of innovation. The dark highlighted area is the average of all states in the index.

The technological possibilities of Industry 4.0 are innumerable, with its limitations arising from human capital. Mařík [4] draws attention to the fact that the fourth industrial revolution is a more fundamental change in human thinking than in technology – we talk here of the term Thinking 4.0. Demands are placed on universities, which must change the content and style of tuition, and this itself imposes certain financial demands. Investment will move to education, where a change in the method of evaluation and management of universities will be required.

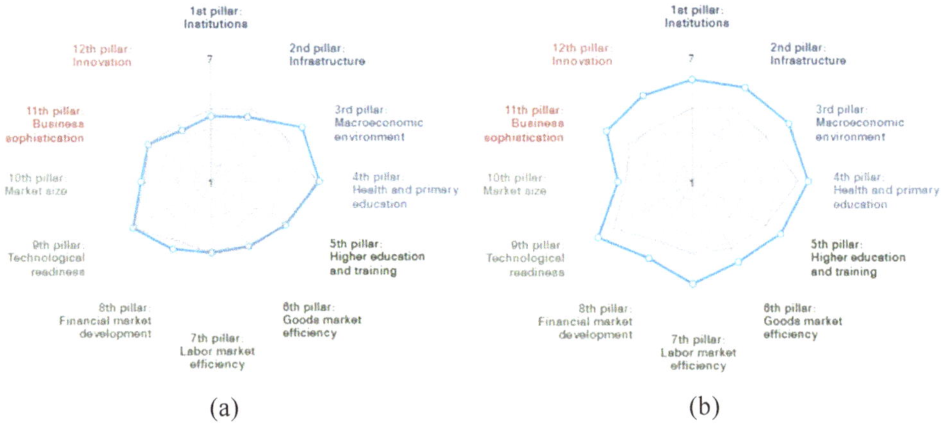

Figure 1: (a) Czech Republic 2017–2018; and (b) Switzerland 2017–2018 [3].

The main focus of this paper is analysis of the current situation on the labour market from the viewpoint of employers and their requirements of students and school leavers in the Czech Republic. This analysis focuses on students at the Faculty of Safety Engineering at VSB – Technical University of Ostrava from the viewpoint of their knowledge of Industry 4.0.

2 QUESTIONNAIRE SURVEY

A quantitative method of public opinion research was used for pilot analysis, i.e. a questionnaire survey with the use of statistical techniques. The data obtained by the questionnaire survey was subjected to three hypotheses tested at the 5% level of significance ($\alpha = 0.05$). A test of goodness of fit (Pearson's qui-squared test), which is used on categorical data for comparing frequencies, was used to test the given hypotheses. The general formulation of the null and alternative hypothesis is as follows in this test:

- H_0: frequencies are equal;
- H_A: frequencies differ significantly.

Equation for calculation of test statistics (1):

$$\chi^2 = \sum_{i=1}^{n} \frac{(O_i - E_i)^2}{E_i}, \tag{1}$$

where

O_i the observed frequency of the *i*th group;
E_i the expected frequency of the *i*th group;
n the number of groups;
χ^2 the qui-squared distribution value of the test statistics.

The value of the test statistics was transformed using the CHIDIST function to the attained value of significance (p-value). This value was subsequently compared with the chosen level of significance (α). The null hypothesis was rejected in favour of the alternative hypothesis if the attained level of significance was lower than the chosen level of significance. The required calculations were performed in the software Microsoft Excel 2007.

2.1 The questionnaire survey of companies in the Czech Republic

The first questionnaire comprised of 11 questions was concerned with research into enterprise entities in the context of Industry 4.0 in the Czech Republic. The aim of the questionnaire was to analyse the current situation at companies, particularly their position in relation to schools and their requirements of graduates. The target group were companies that have already introduced Industry 4.0 technology in order for their feedback to contribute to the future transformation of study fields responding to the fourth industrial revolution. The total number of questionnaires evaluated was 87, of which 56 were from companies that have already introduced the given technology.

2.1.1 Evaluation of the questionnaire

The representation of respondents by size of company was relatively even. The greatest proportion of respondents represented enterprise entities in industry, i.e. the secondary sector, and amounted to 52 respondents. This was followed by the tertiary sector with 33 respondents representing services – transport, the health service, trade, education, cultural and communal services, science and research, high-tech technologies, etc. There was only minimal representation of the primary sector, amounting to just 2 respondents.

More than a half of the 87 respondents completing the questionnaire have already introduced Industry 4.0 technology at their companies. Fig. 2 depicts the technologies most strongly represented.

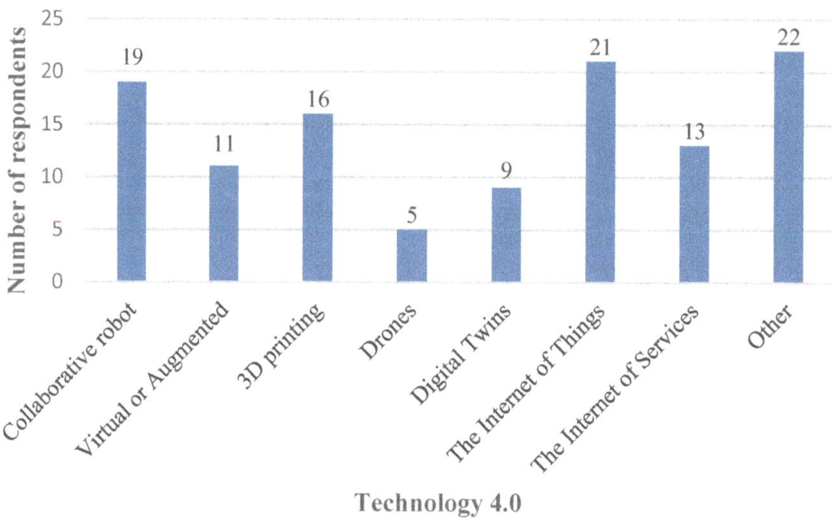

Figure 2: Industry 4.0 technologies represented among the questionnaire respondents.

Most of those questioned, i.e. 82%, give preference to graduates of technical fields associated with work in industry. A total of 74% of respondents were/are in contact with a university within a framework of provision of work experience for students during the course of their studies. A number of companies offer paid work experience, though this is not a condition. The aim is to facilitate contact for students both with work experience and with potential future employment, giving companies the chance of gradually educating their own future employees. The research indicated that 26% of respondents have never engaged in co-operation with a university, i.e. 23 of the surveyed companies. 15 of these 23 surveyed companies would, however, welcome such co-operation in the future.

74% of respondents gave preference to graduates with work experience. Conflict sometimes arises between study and work experience. There are groups of students that actively seek experience in the field during their bachelor studies, and these groups find it simpler to obtain valuable experience. There is then a group of students that devote themselves exclusively to their education during the course of their studies, and leave their initial contact with the world of work until after they have completed their education. In their job advertisements, companies often state the ambivalent words "suitable for graduates" and "we require at least 2 year's experience". One possible solution is greater emphasis on and motivation for the expansion of co-operation between companies and universities.

The respondents expressed rather similar views regarding the knowledge and skills that graduates should have after completing university. 53% of those surveyed saw no lack of knowledge/skills in graduates. The remaining 47% saw a lack of experience in various areas. These areas were analysed and divided into groups. The first group contains knowledge that could be developed by students at university in the future within the scope of their study field: knowledge of the legislation (general safety, ATEX), mechanics, physics and materials, language skills, work experience, conscientiousness, a desire to learn new things, independence and analytical thinking. The second group contains knowledge relating to selected study fields. There is demand for students who have studied IT technologies, electrical engineering and engineering.

Responses to the final question indicate that companies want to invest means most often in 3D printing technology, expanded and virtual reality, and collaborative robotics.

2.1.2 Hypothesis 1: The university at which an applicant has studied is not important to employers

To test this hypothesis, we used the answers given to the question: "Do you give preference to particular universities at which applicants have studied?" The frequencies of the responses are summarised in Table 2.

Table 2: Observed and expected frequencies of responses for hypothesis 1.

	Observed	Expected
Yes	19	43.5
No	68	43.5
TOTAL	**87**	**87**

Hypotheses:

$H_0: n_N = n_Z,$

$H_A: n_N \neq n_Z,$

where n_N represents the frequency of respondents for whom the specific university attended is not important and n_Z represents the frequency of respondents for whom the specific university attended is important.

Results:

$$\chi^2 = \frac{(19-43.5)^2}{43.5} + \frac{(68-43.5)^2}{43.5} = 27.5977$$

$$p - value = 1.49 * 10^{-7}$$

Conclusion:
The attained level of significance ($1.49 * 10^{-7}$) was smaller than the chosen level of significance (0.05). The test showed that the frequency differs significantly; essentially it was demonstrated that the frequency n_N (68) is significantly different, i.e. higher, than frequency n_Z (19), for which reason the examined hypothesis "The university attended is not important to employers" is confirmed at the 5% level of significance.

2.1.3 Hypothesis 2: The knowledge of students is inadequate in relation to the requirements of employers following completion of university

To test this hypothesis, we used the answers given to the question: "Have you ever felt a lack of specific knowledge or skills in employees in the context of your technology?" The frequencies of the responses are summarised in Table 3.

Table 3: Observed and expected frequencies of responses for hypothesis 2.

	Observed	Expected
Adequate	54	43.5
Inadequate	33	43.5
TOTAL	**87**	**87**

Hypotheses:

$H_0: n_N = n_D,$

$H_A: n_N \neq n_D,$

where n_N represents the frequency of respondents who felt a lack of specific knowledge or skills in graduates (they had inadequate knowledge) and n_D represents the frequency of respondents who did not feel a lack of specific knowledge or skills in graduates (they had adequate knowledge).

Results:

$$\chi^2 = \frac{(54 - 43.5)^2}{43.5} + \frac{(33 - 43.5)^2}{43.5} = 5.069$$

$$p - value = 0.0243$$

Conclusion:
The attained level of significance (0.0243) was smaller than the chosen level of significance (0.05). The test showed that the frequencies differ significantly; essentially it was demonstrated that the frequency n_N (33) is significantly different, i.e. lower, than frequency n_D (54). It is clear at first glance from the data that the hypothesis cannot be proven, for which reason our investigated hypothesis that "The knowledge of students following completion of university is inadequate in the context of the demands of employers" was rejected at the 5% level of significance.

2.1.4 Hypothesis 3: More than half of employers surveyed who are not engaged in co-operation with a university within the framework of work experience for students would welcome future co-operation

To test this hypothesis we used the responses to the questions "Have you ever engaged in co-operation with a university within the framework of work experience for students?" and "If not, would you welcome the introduction of co-operation on academic soil?" The frequencies of the responses of respondents who were not engaged in co-operation with a university are summarised in Table 4 along with the expected frequencies of the test of good fit.

Table 4: Observed and expected frequencies of responses for hypothesis 3.

	Observed	Expected
Would welcome	15	11.5
Would not welcome	8	11.5
TOTAL	**23**	**23**

Hypotheses:

$H_0: n_{ANO} = n_{NE},$

$H_A: n_{ANO} \neq n_{NE},$

where n_{ANO} represents the frequency of respondents who were not engaged in co-operation with a university, but would welcome future co-operation, and n_{NE} represents the frequency of respondents who were also not engaged in co-operation with a university, but reject future co-operation with a university.

Results:

$$\chi^2 = \frac{(15 - 11.5)^2}{11.5} + \frac{(8 - 11.5)^2}{11.5} = 2.130$$

$$p - value = 0.144$$

Conclusion:

The attained level of significance (0.144) was higher than the chosen level of significance (0.05), for which reason we do not reject the null hypothesis in favour of the alternative. The investigated hypothesis "More than half of surveyed employers who are not engaged in co-operation with a university within the framework of work experience for students would welcome future co-operation" could not be confirmed at the 5% level of significance.

2.2 Questionnaire survey of students

A second questionnaire investigated the awareness of students of Industry 4.0. The target group were students at the Faculty of Safety Engineering at VSB Technical University of Ostrava. A total of 70 students participated in the questionnaire survey, of which:

- 28 students in the first year of bachelor studies;
- 15 students in the second year of bachelor studies in the study field Safety of Work and Processes;
- 19 students in the third year of bachelor studies in the study field Safety of Work and Processes; and
- 8 students in the first year of master's studies in the study field Safety Engineering.

The questionnaire was comprised of five questions with a maximum score of six points. The boundary assessing success was set as a range of 4–6 points. Questionnaires with three points or less were assessed as unsuccessful. The aim was not the individual assessment of those completing the questionnaire, but merely a general overview of the basic knowledge the students have in the given area.

2.2.1 Assessment of the first year of bachelor studies
Only 13 of the 28 students who completed the questionnaire achieved a successful score. The remaining 15 were counted as unsuccessful. The relative frequency of correct answers in the following graph indicated the lack of knowledge among the students answering the question "Which country is considered the cradle of Industry 4.0". Only three of the 28 students gave the correct answer – Germany. The remaining answers were assessed positively.

2.2.2 Assessment of the second year of bachelor studies
Only 7 of the 15 students who completed the questionnaire achieved a successful score of four points, i.e. less than half. This level of success is, therefore, comparable, with that of students in the first year of bachelor studies.

2.2.3 Assessment of the third year of bachelor studies
The least successful year from the viewpoint of our analysis is the third year of bachelor studies. Only three of the 19 students who completed the questionnaire achieved the successful score of four points. A total of 11 students attained a score of three points in the questions answered. Only 10% of these students completed the questionnaire successfully.

2.2.4 Assessment of the first year of master's studies
The smallest number of students in all the years analysed was seen in the first year of master's studies – eight respondents. Only two of these students attained the successful score of four points. In percentage terms, two successful questionnaires represents just 25%, which is the second worst result of all years, regardless of the small group of those completing the questionnaire.

2.2.5 Summary evaluation
The most successful year in terms of the number of correct answers (a score of five points) is the second year of bachelor studies, followed by the third year. The difference between years two and three is 22%. The first year of bachelor studies, with 46% success (success meaning at least four correct answers on the questionnaire) was ranked in third place. The first year of master's studies achieved the result of 25% with four correct answers, in spite of the low number of respondents. The average number of correct answers can be seen in Table 5.

Table 5: Average number of correct answers.

1st year of bachelor studies	3.25
2nd year of bachelor studies	3.53
3rd year of bachelor studies	2.89
1st year of master's studies	2.75

3 DISCUSSION AND RECOMMENDATIONS

Measures that could contribute towards assuring greater employability of graduates, particularly in Industry 4.0, are proposed in connection with the analyses performed.

3.1 The amendment of curricula in the study programme Fire Protection and Industrial Safety, The Faculty of Safety Engineering at VSB – Technical University of Ostrava

Amendment to the subject Machinery, Equipment and Technology is recommended in the 1st year of bachelor studies, which is the same for all first-year students.

Amendment to the subjects The Safety of Machinery and Equipment, Security Informatics, Technical Safety, The Psychology of Work Safety, and Risk Analysis and Danger is recommended in the 2nd, 3rd and 4th years of bachelor studies in the study field The Safety of Work and Processes.

Amendment to the subjects Diagnostics and Maintenance, and Developmental Trends in Safety Engineering is recommended in the 1st and 2nd years of master's studies in the study field Safety Engineering.

3.1.1 Specific exercises for students in the study field The Safety of Work and Processes

Knowledge of the following legal standards is recommended for students within the framework of understanding the basic issues relating to collaborative robots (cobots):

- ISO 13482:2014 Robots and robotic devices – Safety requirements for personal care robots;
- ISO 15066:2016 Robots and robotic devices – Collaborative robots.

A cobot is more expensive and slower than a "normal" industrial robot, though more suitable for certain applications, even though it does not co-operate with man, because information on the pressure of its tensometers (sensors) can be used. A cobot can be viewed from two perspectives from the viewpoint of its use in practice:

1. The robot is in collaborative mode – in this case the ISO standard 15066:2016 applies, from the table of maximum forces (various forces for various parts of the body) that the robot can exert during contact with humans. Impact with any force to the head and neck is prohibited. If the robot fulfils these parameters, a person may approach to any distance. If the robot does not fulfil these parameters, it is necessary to calculate the safe distance for stoppage, this means
2. The robot in non-collaborative mode – the safe distance is calculated from CSN EN ISO 10218-1:2011 Robots and robotic equipment – Demands for the safety of industrial robots – Part 1: Robots. The minimum distance to the unsafe space must be calculated by the basic eqn (2):

$$S = (K \times T) + C, \tag{2}$$

where

S is the minimum distance (mm);
K the parameter of speed of approach (mm/s);
T the total period of system stoppage (s);
C the distance of penetration (mm).

3.2 Increased awareness among students of Industry 4.0 in the form of lectures and seminars

This issue can also be incorporated into other subjects taught at faculties of VSB – Technical University of Ostrava to increase awareness of the fourth industrial revolution among students and draw attention to the importance of study of this issue. Lectures and seminars may be given by professors and engineers of other universities and by professionals owning and using 4.0 technology. In general terms, it is recommended that experts, company owners and company employees are engaged in education to a greater extent. The kind of lectures they give and their practical experience are extremely valuable and motivational for students, and this form of tuition operates in certain foreign countries.

3.3 Introduce intensive co-operation with the Czech Institute of Informatics, Robotics and Cybernetics (CIIRC), which is part of the Czech Technical University in Prague

CIIRC is a research centre and workplace in industry that is a highly regarded institute in the Czech Republic within and beyond the academic sphere. It also serves the students who visit it during their studies and who can try out practical procedures and technology here that will be, or already have been, introduced into industry. Part of CIIRC is a Testbed for Industry 4.0, which serves as a platform for contact between companies and the academic sphere. VSB – Technical University of Ostrava is also to have its own Testbed for Industry 4.0 in the near future, and the golden rule that the most effective way is to learn from the very best certainly continues to apply here.

3.4 Do more to address companies from all over the Czech Republic and introduce co-operation for the purpose of work experience for students during the course of their studies

Both the university and its individual faculties and students who can arrange co-operation near their homes can help expand this form of co-operation. Students can choose companies by either sector, size or principal area of business. A great way of repaying a company's loyalty is to invite it to the annual jobs fair Career PLUS where awareness of companies can be spread among the general public and where they can come into contact with students and the academic sphere.

4 CONCLUSION

This paper focuses on analysing the current situation on the labour market from the viewpoint of employers and their requirements of students and university graduates in the Czech Republic in connection with Industry 4.0. A questionnaire survey of companies, whose respondents were employers in various branches, was evaluated. The data obtained was subjected to three hypotheses tested at the 5% level of significance. A test of good fit (Pearson's qui-squared test) was used to test these hypotheses. The awareness of Industry 4.0 among students at the Faculty of Safety Engineering at VSB – Technical University of Ostrava was analysed, and recommendations that may contribute towards assuring greater employability of graduates, particularly in the area of Industry 4.0, were proposed on the basis of the results of the questionnaire survey.

The introduction of trends of Industry 4.0, competitiveness, innovation, digitalisation and the creation of new job opportunities is conditional, to an ever increasing extent, on the use of new information and communication technologies (ICT). It is important that the population has the necessary knowledge and skills associated with the use of ICT. Their broader and more intensive use leads to growing demand for professionals who are capable of performing the great quantity of tasks associated with the development and maintenance of corporate ICT systems. In the Czech Republic, as in other countries, we have seen a trend in recent years towards an increasing proportion of concerns preferring to purchase ICT services rather than establish their own ICT departments. According to Eurostat research [5], concerns most frequently purchase activities related to the maintenance of ICT infrastructure

(58%), followed by activities associated with data security and protection (53%) and the development and maintenance of webpages (49%).

There is growing demand on the labour market for experts in the compilation, programming and administration of new technologies, business and data analysts, ICT project managers, ICT product dealers, security specialists and IT risk analysts, developers of web and mobile applications, software and graphics for the purposes of entertainment, experts in the development and support of cloud computing services, and embedded systems programmers. In the development of IT interfaces for human–machine work, for example, it will be necessary to have not just IT and technical knowledge, but also knowledge of, for example, ergonomics and the prevention of health risks. It can be said, in general terms, that the demand from companies for ICT professions will also increasingly include the demand and need for soft skills, enthusiasm for the field, communication skills, independence, initiative, the ability and willingness for continual self-education, flexibility, and logical and analytical thought. Higher positions will demand, first and foremost, managerial skills and creativity. It can be said in summary that companies expect of universities graduates with a multidisciplinary education who are creative, innovative and independent and, most importantly, who will have greater practical experience of the real working world.

Companies approach the employment of university graduates in two ways. The first seek suitable candidates for required positions on the labour market. An active approach is chosen, first and foremost, by larger companies who seek out their potential future employees among students with a suitable professional orientation. They often have a sophisticated training programme for students, offer work experience and provide suggestions for diploma subjects and guidance for students in co-operation with selected universities, or offer the chance of education for university students in a special education centre. They test potential job applicants in this way. This work experience is applied, first and foremost, by companies with foreign owners who have a sophisticated system of work with interns that is generally an adaptation of the system used at the parent company. Companies with Czech owners often complain of the fact that, given the shortage of graduates of ICT fields and the great demand for these professions on the labour market, they can compete with companies with foreign owners only with considerable difficulty in the acquisition of young professionals. Smaller companies give precedence to people with work experience when it comes to IT professions.

ACKNOWLEDGEMENT

The authors gratefully acknowledge the financial support of the project "Potential Impacts of Industry 4.0 on Operators in 3.0 Jobs and Tertiary Education in Accordance with Safety Engineering", identification code TL01000470.

REFERENCES

[1] Sankot, O., Global Competitiveness Report 2016–2017: Management summary. Competitiveness CZE 2017 – Digital Society, 2017. https://fmv.vse.cz/wp-content/uploads/2016/09/manazerske_shrnuti_GCR_201617.pdf. Accessed on: 25 Feb. 2019. (In Czech.)

[2] Schwab, K. & Sala-i-Martin, X., The Global Competitiveness Report 2016–2017. World Economic Forum, 2017. www3.weforum.org/docs/GCR20162017/05FullReport/TheGlobalCompetitivenessReport2016-2017_FINAL.pdf. Accessed on: 25 Feb. 2019.

[3] Schwab, K. & Sala-i-Martin, X., The Global Competitiveness Report 2017–2018. World Economic Forum, 2018. www.weforum.org/reports/the-global-competitiveness-report-2017-2018. Accessed on: 25 Feb. 2019.

[4] Mařík, V., *What is Meant by Industrie 4.0 and Industry 4.0?* Czech Institute of Informatics, Robotics, and Cybernetics (CIIRC), Czech Technical University in Prague, 2016.
[5] Eurostat, http://ec.europa.eu/eurostat/statistics-explained/index.php/ICT_specialists_-_statistics_on_hard-tofill_vacancies_in_enterprises. Accessed on: 25 May 2019.

TRAINING EFFECTIVENESS ANALYSIS OF OSHA SILICA AND EXCAVATION STANDARDS FOR CONSTRUCTION

EMRAH KAZAN, MUMTAZ USMEN, BEDEL DESRUISSEAUX,
SERHAN KAYA & MUSSIE SEYOUM
Civil and Environmental Engineering Department, Wayne State University, USA

ABSTRACT

As construction safety and health standards evolve due to changing industry practices and stakeholder expectations, and the mandates get stronger on compliance with these standards, the need for training programs become more pronounced. This paper covers the training material development and program delivery and evaluation efforts associated with two standards promulgated by the US Department of Labor Occupational Safety and Health Administration (OSHA), namely Respirable Crystalline Silica (29 CFR 1926.1153) and Excavation and Trenching Safety (29 CFR 1926 Subpart P). The work reported here was undertaken through OSHA's Susan Harwood program. The developed training materials consisted of PowerPoint instructional modules; pre-tests and post-tests to measure incremental knowledge improvement; and trainee opinion surveys to evaluate the efficacy of the training materials and training delivery systems incorporated in the programs. Trainees included employees and employers representing various trades (operating engineers, laborers, masons, bricklayers, pipelayers and others). The delivery was performed by a traditional classroom format for the silica training, while lecture-based traditional and online training were separately offered for excavation safety. Kirkpatrick's model for learning assessment was adopted, employing level 1 and level 2 evaluations. It was found that both the lecture-based traditional and online training delivery were effective tools, although they have their advantages and disadvantages. Based on a limited study using cross-tabulation, no statistically significant correlation could be identified between levels 1 and 2 assessment results.

Keywords: training effectiveness, Kirkpatrick model, OSHA standards, respirable crystalline silica, excavation and trenching safety, traditional and online training.

1 INTRODUCTION

Construction is recognized as a high hazard industry in the US, as well as globally. A wide range of construction tasks and activities expose workers to physical dangers and/or health hazards, and if safe work practices are not followed, the consequences can be severe and costly. In the US alone, the annual total costs of fatal and non-fatal injuries in the construction industry is estimated to be over $10 billion. Research and practice have shown that effective safety training is a critical part of achieving workplace safety. This can enhance the workers' recognition and understanding of potential hazards, and improve their safety responses accordingly to avoid injury, illness and fatality.

According to the annual data published by the Bureau of Labor Statistics (BLS), there were 194,300 injuries and 3,800 illnesses reported for the construction industry in 2018. Although there have been improvements in these statistics over the past few decades, these numbers are still considered high and unacceptable. On the other hand, 971 construction worker fatalities were recorded in 2017, and this number has been persistently high over the over the same period, reflecting critical needs to take corrective action. Broader adoption of company health and safety programs by the private construction sector with an effective training component is considered essential to improve the safety culture and reduce injuries, illnesses and fatalities in order to provide workplaces that are free of hazards.

The Occupational Safety and Health Administration (OSHA) promulgates and enforces safety and health standards applicable to all industries; however, it is noteworthy that there

www.witpress.com, ISSN 1743-3509 (on-line)
doi:10.2495/SAFE190041

are a set of standards specifically applicable to construction (29 CHR 1926) because of its unique characteristics. Due to factors such as the fragmentation of project participants, large forces involved in constructing the structures using various equipment and tools, highly variable and sometimes difficult-to-control site conditions, and the transient/seasonal nature of the workforce, construction is viewed as an area requiring special focus [1]. Besides site inspections resulting in citations and penalties for safety violations, OSHA has also developed and delivered technical support and training programs to reach out to the industry with the objective of increasing the safety awareness of the workforce in critical high hazard areas, while aiming to impart better knowledge and skills needed for safer operations. The Susan Harwood program is one such initiative that competitively avails funding to non-profit agencies and educational institutions on an annual basis. Funding awardees develop and implement training programs on targeted topics to workers in industry sectors experiencing high injury, illness and fatality. Typically, the scope of work under the grant covers training material development, training delivery, and training effectiveness evaluation. For each grant, there is an expectation that the grantee trains a minimum number of eligible personnel on selected topic(s) and reports the number of training hours completed in the process. Approval of the training materials and delivery systems, grant monitoring and training observations are OSHA's responsibility, and the grant team submits quarterly and close-out reports to OSHA for review, feedback and approval. In most cases, the grantees collaborate with industry training experts and leaders by forming an advisory committee that provides guidance and feedback to the project team, while also assisting with the identification and recruitment of the trainees.

We have received and completed two Susan Harwood grants in the past few years; one on Respirable Crystalline Silica training and another on Excavation and Trenching Safety training. The impetus for OSHA to solicit training grant proposals for silica training through the Susan Harwood program was the continuing health concerns regarding the exposure of construction workers to excessive amounts of silica dust (possibly leading to silicosis and lung cancer), prompting it to issue a new standard (29 CFR 1926.1153) and making relevant training resources available to employees and employers nationwide, so useful information and knowledge on the contents of the standard (29 CFR 1926.1153) could be disseminated to appropriate groups and individuals to facilitate better compliance. Our second training grant was on excavation and trenching safety, which was funded by Susan Harwood especially because the fatality statistics on trench cave-ins and other serious excavation related accidents called for continuation and reinforcement of transfer of knowledge on the OSHA standard to targeted audiences. We were able to train a total of 721 people on silica in 2,163 contact hours, and 405 people on excavation safety in 1,215 contact hours through these grant efforts. While implementing these activities, we collected a considerable amount of demographic and knowledge gain and improvement data, and were able to analyse this data to learn about training effectiveness evaluation. This paper presents the results of these analyses and draws conclusions on the approaches that can be taken for training delivery and their effectiveness.

2 OBJECTIVES AND APPROACH

Given the mission of the Susan Harwood program, the main objective of our grants were to identify and train a minimum number of eligible personnel in the construction industry while achieving a minimum threshold of training contact hours (excluding preparation and travel). In accordance with our proposals, we committed to prepare PowerPoint training modules on the targeted topics; to present these modules to targeted audiences that need to be trained;

and to evaluate the effectiveness of the training materials, delivery and the overall program. The tasks performed by our research team included the following:

- Preparation of the training materials, including PowerPoint modules (consisting of text, graphics, photos and videos), a pre-test and post-test(s) measuring the knowledge gain/improvement from the instruction, and an opinion survey questionnaire to capture demographic information and record the perceptions of the trainees regarding the efficacy and effectiveness of the training they received.
- Delivery of the training content (via PowerPoint slides) to groups or individuals categorized as employees or employers, representing construction trades, such as operating engineers, laborers, masons, bricklayers, and others using traditional classroom lectures (with demonstrations as needed), or by allowing them to take the same training online, facilitated by an internet portal developed by our team.
- Collection of quantitative and qualitative data from the trainees along the grant period, combined with feedback from the advisors and OSHA experts reviewing our work, to continuously assess, revise and improve the materials and delivery system.

The grant period was one year for each of the projects. The training sessions using the traditional format took place in venues such as the trade organizations' or contractors' training centers, or other venues provided by the contractors (e.g. warehouse facilities). A typical training session took three hours and followed the flow diagram presented in Fig. 1. The steps in a given session include trainee sign-up; pre-test; PowerPoint presentation (with questions, answers and discussions); post-test (s); and opinion survey. It should be noted that although typically, there is one pre-test at the beginning of the session after sign-up, there may be multiple post-tests given after completing some part of the module. The tests are graded on-site, and the correct answers are shared with the audience, with additional discussions as appropriate to augment learning. A certificate is provided to the trainees to indicate their participation in or completion of the particular training. These certificates are mailed to the trainees.

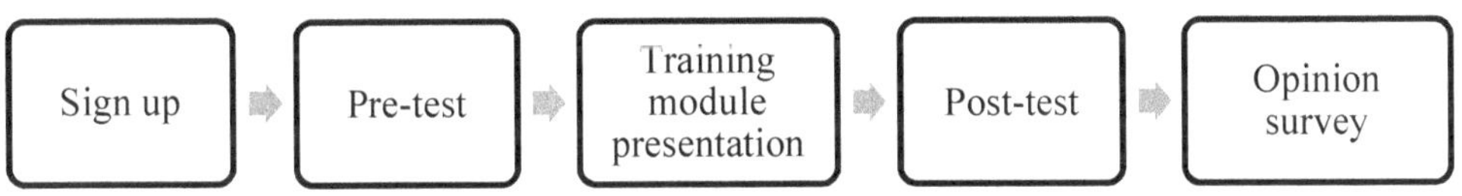

Figure 1: Typical training flow chart.

The online format follows the same logic; however, it is self-paced and no opportunities for any interaction or discussion online are currently available. The portal is accessed by a universal resource locator (URL) link, or by scanning a Quick Response (QR) code (see Fig. 2 for a sample), which is suitable for handheld devices such as cell phones and tablets. This format of delivery has the distinct advantage of flexibility in that it provides on-demand training that can be taken anytime and anywhere at the trainees' convenience. The portal is designed to capture essentially the same type of research data, and has the capability to post the grades earned in the pre-test and post-test(s), while also displaying the correct answers to each question after the test is finalized. A certificate can be downloaded from the computer, laptop, tablet or smart phone upon completion of the training session. There is no time limit.

Figure 2: Sample QR code.

Table 1: Training module outlines.

Silica module	Excavation module

Introduction
• Respirable crystalline silica and silica exposures
• Silica exposure associated health risks
New OSHA standard on respirable crystalline silica
• Permissible exposure limit and action level for construction
• Scope and definitions
• Compliance dates
Silica exposure control methods for construction
• Specified exposure control methods (Table 1 of standard)
  ◦ Engineering controls
  ◦ Workplace controls
• Alternative exposure control methods
  ◦ Exposure assessment and reassessment
  ◦ Performance and scheduled monitoring options
  ◦ Methods of sample analysis
  ◦ Employee notification
  ◦ Hierarchy of controls
  ◦ Abrasive blasting
Mandatory requirements for silica exposure control
• Respiratory protection
• Housekeeping
• Written exposure control plan and competent person
• Medical surveillance
• HazCom-silica
• Record keeping
Compliance summary
	Introduction
	• Excavations and trenches
	• How cave-ins occur
	• How cave-ins cause fatality
	• Excavation and trenching fatality and injury statistics
	• Other excavation and trenching hazards
	Soil classification
	• Categories (solid rock, Type A, Type B, Type C soils)
	• Soil testing
	• Downgrading soil category
	• Water effect on classification
	Cave-in protective systems
	• Engineering controls
	  ◦ Protective systems
	  ◦ Sloping, benching, shoring
	  ◦ Access and egress
	  ◦ Installing/removing protective systems
	  ◦ Protective system maintenance
	• Workplace controls
	  ◦ Training
	  ◦ Site inspection, competent person, P.E.
	Other excavation hazards
	• Struck-by hazards
	• Fall hazards
	• Underground/overhead utilities
	• Hazardous atmospheres
	• Respiratory protection/ventilation
	Rescue operations
	Safe practices checklist

3 TRAINING EFFECTIVENESS EVALUATION METHODOLOGY

Evaluation of learning/training effectiveness is essential for gauging the success of a training program. Limited information is available in the literature on this topic. The relevant papers include: Kazan and Usmen [2], Impact Measurement Centre [3], McCann [4], Tamkin et al. [5] and Wang and Wilcox [6]. As discussed by Tamkin et al. [5], Donald Kirkpatrick offers a model for training effectiveness evaluation, which refers to four levels of assessment: Level 1 assessment is the reaction of the trainee to the perceived usefulness of the training received. Level 2 is the learning increment evaluated by comparison of the scores of a pre-test given prior to the PowerPoint instructional module and the post-test(s) covering all or a part of training lecture(s) presented. Level 3 assesses the retention of knowledge and skill gained in initial training as determined by evaluating the trainees' safety behavior after a certain period has elapsed following the initial training. Level 4 is the actual result/impact (in terms of site/company safety improvements) over a period of time using a set of leading and lagging indicators of safety as assessment metrics. In our grant work, we used only Level 1 (derived from the survey) and Level 2 (based on pre-test and post-test scores).

For level 1, the following questions (presented as statements) were included in the opinion survey using a 5-point Likert Scale (1 – strongly disagree, 2 – disagree, 3 – neutral, 4 – agree, 5 – strongly agree):

- SQ1. This safety training improved my knowledge. (PKI – Perceived knowledge improvement)
- SQ2. Overall, the safety training materials presented were easy to understand.
- SQ3. The topics covered was relevant to me.
- SQ4. This training experience will be useful in my work.

For Level 2, the metric for knowledge gain/improvement as a result of training was formulated as

$$\mathrm{KIR} = \frac{\mathrm{POT}}{\mathrm{PRT}}, \tag{1}$$

where KIR is the knowledge improvement ratio; POT is the post-test result, and PRT is the pre-test result.

The KIR value as originally conceived by Kaya [7] is helpful not only in evaluating the amount of incremental knowledge gain, but also in identifying further training or retraining needed by program participants. KIR greater than 1 illustrates knowledge improvement; KIR less than one implies knowledge regression; KIR of one means no change in knowledge level.

4 TRAINING EFFECTIVENESS DATA ANALYSIS

The collected data was statistically analysed using the Statistical Package for the Social Sciences (SPSS) software. The results are presented in the following paragraphs.

4.1 Trainee demographics

The silica training program which reached out to 721 personnel, it was found that 48.5% of the trainees had high school education/GED; and 31.5% had some college; only 2.5% lacked a high school diploma. Operating engineers constituted 50% of the program participants, followed by masons (11%), laborers (10%), and bricklayers (9.5%). Majority of the trainees (52%) had more than 20 years of experience, and a small percentage (7.4%) had less than 10 years of experience. A total of 72% of the trainees reported that they had prior safety training.

A total of 405 personnel were trained in the excavation training program, however, data was available for only 176 responses related to education level. Based on this sample, 49% of the trainees reported to have high school education/GED; 31% had some college; and bachelor and associate degree holders were 4% each. The rest of the questions received responses from all 405 participants. Referring to this data, operating engineers populated 39% of the program; 19% was construction laborers, and 17% were pipelayers. Majority of the trainees (55%) had 1 to 5 years of experience; 18% had between 6 and 15 years; while about 19% had more than 20 years of experience. About 58% of the trainees had prior safety training.

4.2 Knowledge improvement

The results of knowledge improvement analyses for silica and excavation training programs are summarized in Table 2. It was determined that satisfactory performance for any trainee would be achieving a score of 72 (18 correct answers on the pre-test or post-test out of 25 questions, expressed as a percentage). The data can be analysed under four different scenarios as delineated in the table. The separation of the data into four groups is necessitated by the fact that KIR is a ratio while the satisfactory test score is 72. This means that for very high or very low KIR values there may be extreme situations making the interpretation of results unrealistic.

Table 2: Training knowledge improvement data.

Scenarios	Knowledge Improvement Ratio (KIR)	Number of trainees		
		Silica – Traditional	Excavation – Traditional	Excavation – Online
1) Pre-test greater than or equal to 72 and Post-test greater than or equal to 72	< 1	21	2	1
	= 1	38	6	7
	> 1	188	92	89
2) Pre-test greater than or equal to 72 and Post-test less than 72	< 1	8	0	2
3) Pre-test less than 72 and Post-test greater than or equal to 72	> 1	363	68	117
4) Pre-test less than 72 and Post-test less than 72	< 1	8	0	4
	= 1	19	1	1
	> 1	76	7	8
	Total	721	176	229

Referring to Table 2 for data pertaining to silica, Scenario 1 depicts the cases where the trainees scored 72 or higher in both the pre-test and the post-test, which represents a total of 247 cases out of 721, or 34% of the total population. Interestingly, there were 38 cases with no knowledge improvement (KIR = 1) and 21 cases with knowledge decline (KIR < 1), although the trainees in the latter case still successfully completed the entire training program. There was a very small number of trainees (8) in Scenario 2, representing the people who regressed from a satisfactory to an unsatisfactory score. In this case, the trainee would be considered unsuccessful based on the post-test score, even though s/he could have been deemed successful just based on the pre-test. Scenario 3 is the most impactful case, with KIR greater than 1, where the trainees increased their knowledge from an unsatisfactory (post-test score below 72) to a satisfactory level (above 72). The number of people in this category was 363, corresponding to 50 percent of the total population who received silica training. Finally, in Scenario 4 the trainees failed both the pre-test and the post-test; a positive outcome in this case was that 76 trainees (11%) improved their knowledge, however this occurred without achieving a satisfactory score.

The same four scenarios also appear in Table 2 for excavation training, which was carried out with lecture-based traditional format and using the online delivery, thus allowing for comparisons. It appears from the data shown in the table that people who received traditional training did better than those who completed online training in Scenario 1. That is, 92 trainees out of 176 total (55%) demonstrated knowledge improvement when both the pre-test and post-test scores were above 72, in comparison to 89 out of 229 people (39%) showing knowledge improvement when trained online. The numbers in Scenarios 2 and 4 are too small to make any reasonable interpretations. The impactful Scenario 3 for excavation, where 185 out of 405 (46%) trainees experienced knowledge improvement, scoring higher than 72 in pre-test as well as the post-test, offers another interesting observation. While 68 out of 176 trainees (39%) exhibited knowledge gain with the traditional training experience, a higher percentage, 55% (117 out of 229) achieved similar success when training online, suggesting that online training can be more effective than the traditional one, when the trainees start at a lower than satisfactory knowledge level.

4.3 Perceived knowledge improvement and satisfaction with training

Trainee responses to the four key questions on the opinion survey are displayed in Table 3. Note that the data shown pertains to the responses comprising Scenario 3, where the KIR's are greater than one while the pre-test scores are less than 72. Only the responses for strongly agree or agree (sum of the two) were included in the analysis. Percentage numbers, calculated as the averages of all responses in Scenario 3, encompass traditional silica training, traditional excavation training and online excavation training. It is clear from these numbers that, with one exception of 62% for online excavation training, more than two-thirds of the trainees agreed or strongly agreed with all of the four statements given in the survey questions, indicating rather strong satisfaction with their learning, as well as training experience.

Table 3: Percentages of responses to key survey questions for Scenario 3.

Survey questions	Agree and strongly agree responses to survey questions (%)		
	Traditional silica training	Traditional excavation training	Online excavation training
SQ1	85	91	74
SQ2	72	90	76
SQ3	77	79	62
SQ4	79	88	69

SQ1 was aimed at measuring knowledge improvement as perceived by the trainees after completing the training. The percentages presented in the table show that traditional excavation training achieved the strongest favorable response (91%), followed by traditional silica training and online excavation training. SQ2 was about the ease of understanding of the training materials by the trainees. Again, traditional excavation training topped the responses at 90%; however, for this question online excavation training drew better responses than traditional silica training. SQ3 addressed the relevance of the topic to the trainees. While the traditional format was quite well received by the trainees for silica and excavation, the online excavation training experience produced mixed results for this question. Finally, SQ4 explored the trainees' perception on how much they thought the training received would be useful in their work. Once more, the responses from the online excavation training group were relatively weaker than the other two.

Overall, it is fair to state that both training programs were very well received by the trainees, and in the case of traditional excavation training, the responses were overwhelmingly positive. Online training, with one exception, produced the weakest responses overall; it is believed that this may be due to the open/unsupervised participation by the trainees whose backgrounds and motivations might vary. This aspect needs further investigation. Comparing silica with excavation, it seems that the latter got higher (better) percentages for all four survey questions. This is more likely because the OSHA standard on excavation and trenching safety has been in effect for several decades, whereas the OSHA silica standard is brand new. Mostly, all personnel receiving the silica training were being exposed to the contents as a first time experience.

4.4 KIR vs. PKI relationship

An attempt was made, as a limited side study, to analyse and determine if there was a possible correlation between KIR and PKI. Cross tabulation analysis was performed with the data (100 trainees) drawn from Scenario 1. It was not possible to do cross-tabulation analysis for Scenarios 2 and 3 because of the fact that KIR is a constant in these cases; it was also unfeasible to perform crosstab analysis on Scenario 4 because of the very small amount of data available. Results for Scenario 1 are given in Table 4. The Chi-square and p (significance) values included in the table were computed by SPSS. Based on these results, it was not possible to obtain a statistically significant correlation between the two variables, meaning that PKI and KIR are independent from each other and should be interpreted individually. This observation indicates that even though trainees may perceive that they increased their knowledge, they may not show this improvement in their post-test scores; vice versa is also true. There have been studies investigating the factors affecting the existence of correlations between Level 1 and Level 2 learning effectiveness [8], [9]. It has

Table 4: Crosstab analysis of KIR vs. PKI for Scenario 1.

Perceived Knowledge Improvement (PKI)	Knowledge Improvement Ratio (KIR)								
	Traditional silica training			Traditional excavation training			Online excavation training		
	<1	=1	>1	<1	=1	>1	<1	=1	>1
Strongly disagree	0	1	5	0	0	0	0	0	4
Disagree	1	1	4	0	1	3	0	0	0
Neutral	1	3	25	0	2	8	0	0	17
Agree	9	14	42	1	1	38	1	3	49
Strongly agree	10	19	112	1	2	43	0	4	19
Total	21	38	188	2	6	92	1	7	89
χ^2 and p-value	$\chi^2 = 7.11$, p = 0.525			$\chi^2 = 7.36$, p = 0.288			$\chi^2 = 6.19$, p = 0.402		

been found that correlations may or may not be strong depending on various factors most of which are trainee-related, e.g. culture, gender, education level. Such an effort was outside the scope of the work reported in this paper.

5 CONCLUSIONS

It can be concluded that knowledge improvements on safety and health standards and their implementation can be realized by training construction trades and other personnel by the traditional and online delivery systems. Knowledge improvement can be quantified by the Knowledge Improvement Ratio (KIR) metric, which is calculated by dividing the post-test score by the pre-test score of a trainee participating in a training session. Another way to get an indication of knowledge improvement is to have the trainees express their perceptions of whether or not their knowledge has improved as a result of completing a training program (PKI), which is usually determined from an opinion survey given at the end of a training session. Online training can follow the same steps as the traditional and has the distinct advantage of reaching out to individuals in remote locations at their convenience in terms of time and place, while generally producing cost savings. On the other hand, traditional training is likely to achieve better knowledge improvement because of the ability of the trainees to interact with the trainers in the course of training and receive answers to their questions with explanations. It is possible, however, to build interactive features into online training, and this is highly recommended. Based on a limited study using Scenario 1 data, it was found that there was no statistically significant correlation between KIR and PKI. We recommend that further study be performed on this relationship, and additionally, the Kirkpatrick learning assessment model be extended into level 3 and level for 4 evaluations to further analyse training effectiveness.

ACKNOWLEDGEMENTS

The funding provided by the US Department of Labor, through OSHA's Susan Harwood Training Program for grants SH-29666-SH6 (silica) and SH-31218-SH7 (excavation) to make this study possible is gratefully acknowledged. The findings of this study reflect the views of the authors, and not those of the sponsor, The US Department of Labor, OSHA.

REFERENCES

[1] National Occupational Research Agenda For Construction, www.cdc.gov/nora/councils/const/pdfs/National_Occupational_Research_Agenda_for_Construction_June_2018_508.pdf. Accessed on: 3 May 2019.

[2] Kazan, E. & Usmen, M.A., Worker safety and injury severity analysis of earthmoving equipment accidents. *Journal of Safety Research*, **65**, pp. 73–81, 2018.

[3] Impact Measurement Centre, Measuring the impact of training and development in the workplace. www.researchgate.net/profile/Anantjyoti_Badgaiyan/post/What_is_a_good_questionnaire_on_Impact_of_training_on_employee_performance/attachment/59d6213a79197b807797fa11/AS%3A295353002807325%401447429052227/download/Measuring_the_Impact_-_Final_Report.pdf. Accessed on: 23 Apr. 2019.

[4] McCann, M., Heavy equipment and truck-related deaths on excavation work sites. *Journal of Safety Research*, **37**(5), pp. 511–517, 2006.

[5] Tamkin, P., Yarnall, J. & Kerrin, M., Kirkpatrick and beyond: A review of models of training evaluation. Institute for Employment Studies: Brighton, UK. www.employment-studies.co.uk/system/files/resources/files/392.pdf. Accessed on: 23 Apr. 2019.

[6] Wang, G.G. & Wilcox, D., Training evaluation: Knowing more than is practiced. *Advances in Developing Human Resources*, **8**(4), pp. 528–539, 2006.
[7] Kaya, S., Effect of construction worker demographics on e-learning with hazard communication standard. Thesis, Wayne State University. https://digitalcommons.wayne.edu/oa_theses/381. Accessed on: 16 May 2019.
[8] Ziegler, B. & Montplaisir, L., Student perceived and determined knowledge of biology concepts in an upper-level biology course. *CBE—Life Sciences Education*, **13**(2), pp. 322–330, 2014.
[9] Grant, D.M., Malloy, A.D. & Murphy, M.C., A comparison of student perceptions of their computer skills to their actual abilities. *Journal of Information Technology Education*, **8**, pp. 141–160, 2009.

SAFETY-ORIENTED MULTI-TASK APPROACH TO MANAGE WHOLE-BODY VIBRATION EXPOSURE AMONG QUARRY OPERATORS

GUIDO ALFARO DEGAN[1], ANDRE ANTONUCCI[2], GIANLUCA COLTRINARI[1] & DARIO LIPPIELLO[1]
[1]Department of Engineering, University of Roma Tre, Italy
[2]University of Chieti G. d'Annunzio, Italy

ABSTRACT

The oscillation generated by external and internal forces is one of the principal characteristics of machinery. The impulses are transmitted to workers' bodies through the part in contact with the vibrational surface, such as the steering wheel of a machine (known as "hand arm vibration" (HAV)), contact surface or the driver's seat of a machine (known as "whole-body vibration" (WBV)). These represent an occupational issue that can generate adverse health effects, annoyance or simple discomfort for the workers during their daily activities. Biodynamic responses of the human body to WBV are reported in terms of the driving point mechanical impedance and apparent mass, transmissibility and the power absorbed by the body. The most common WBV occupational exposure is represented by the operator seated in a variety of vehicle categories such as cars, buses, forklifts, tractors, trucks, and heavy machinery either on or off paved roads in different work contexts. Previous studies investigated the power absorbed by the human body during daily exposure and the consequent potential negative effects on the human health. They evidenced that health effects are connected to the intensity of the vibration levels. In order to reduce workers' exposure, an efficient method could consist of analysing the power frequency spectrum. In fact it provides relevant indications in order to adopt the correct preventative or corrective measurements such as specific suspended seats, extra breaks and individual work schedules. This study is aimed at assessing the WBV exposure among drivers of machinery in a quarrying context thus proving a useful tool to manage such a risk in terms of work organization and scheduling.

Keywords: whole-body vibration, quarrying activities, working task organization, earth moving machinery.

1 INTRODUCTION

Whole-body vibration (WBV) may generate different effects depending on the characteristics of the physical agent itself. Exposure to such an agent may, in fact, generate adverse effects on human health [1] but, in the opposite way, it may induce a good response in the rehabilitation of some chronic diseases as well as for muscle enhancement [2]. However the most common effect is recorded in the field of occupational safety where it represents a risk factor that may affect operators such as drivers of fork-lift trucks [3], tractors [4], loaders [5], LHD vehicles [6]. It's widely recognized that long term exposure to high amplitude WBV is strongly linked to the subsequent development of a particular stress for the human body and back disease has been shown as a connected effect [7]. This last may be influenced with many factors, such as time exposure to the driving task [8], wrong posture [9] or other related factors such as performing manual handling as part of the task during the daily activities [10]. These health problems are very relevant in many workplaces; for example the musculoskeletal disorders (MSDs) accounted for 32% of all injuries and illnesses in 2014 among full-time quarry operators [11], [12]. An accurate risk assessment procedure, based on standards or established norms, should be carried out in order to prevent the negative health outcomes associated with WBV exposure. The most widely accepted standards for measurement and evaluation of human exposure to WBV are the ISO 2631-1 report and the

WIT Transactions on The Built Environment, Vol 189, © 2019 WIT Press
www.witpress.com, ISSN 1743-3509 (on-line)
doi:10.2495/SAFE190051

European Union Directive 2002/44/EC [13]. These norms define specific methods to assess frequency-weighted acceleration amplitude. The use of a method rather than another one is being discussed by many researchers also in the quarry sector [14], [15]. In fact after the publication of ISO 2631-5:2004, both ISO 2631-1:1997 and ISO 2631-5:2004 were utilised simultaneously in research works to assess occupational risks [16]. This feature caused the discussion about the health guidance caution zones (HGCZs) in these two standards. Many researchers have studied these problems regarding of the employment of various vehicles in the quarry activities with particular attention to haul trucks because of their extensive use in the surface mining industry [17]. Even if the reduced number of field tests prevents any strong conclusion, it appears that the different phases of the LHD (Load–Haul–Dump) cycle are different in terms of effects induced among workers [18]. Moreover vibration amplitudes were greatest in the vertical direction, and particular while the trucks were driven unloaded. Another relevant aspect is clearly connected with the sample duration. Eger et al. [19] collected one hour duration WBV measurements from eight haul trucks of varying capacities (30–150 ton) during normal operation at surface mines in Northern Ontario. Seven of the eight measurements were found to lie within the HGCZ defined by ISO2631.1 for r.m.s. and all eight VDV measurements lay within the HGCZ. There is limited evidence to suggest that haul trucks used in surface mining may expose operators to levels of WBV which are associated with risks to the operators' health, however there is some inconsistency in previous reports. These differences are likely to arise from the short measurement durations associated with some of the previous investigations, as well as variability in equipment and sites. During the exposure to the WBV, a driver can be exposed to particular mechanical stress on the body due to vibration peaks. These can be generated by many factors, for example the ground conditions, the vehicle speed and wear degree of the seat. The vehicle constitutes the source vibration and the primary contact point for transmission of vertical WBV to drivers is through vehicle seat (amplifies or attenuates vibration at the base) [20]. Seating system plays a vital role in isolating the vibration. In order to obtain the vibration attenuation, this target may be achieved at different levels. The first one concerns the source vibration, the vehicle. Some wheel loaders employed in the quarry sites are equipped with ride control systems which reduce the vehicle pitching motion [18] and thereby reduce the fore-and-aft vibration exposure, although no study has been made to quantify their efficiency. Sure enough hydro-pneumatic suspensions (HPS), owing to their compact design and superior design flexibility, are increasingly being implemented in commercial road and off-road vehicles [21]. Their nonlinear pneumatic-stiffness and hydraulic-damping properties could provide enhanced attenuation of ride vibrations under large payload variations, while offering improved handling performance through cross-axle interconnections or control interventions. The variations in fluid compressibility and free gas volume cause increase in effective stiffness but considerable reduction in the damping in a highly nonlinear manner. Increasing the gas volume fraction resulted in substantial hysteresis in the force–deflection and force–velocity characteristics of the strut. The second level at which attenuation occurs is at the seat level that represents the contact point between the driver and the vehicle. Although fore-and-aft suspended seats are available on the market, most of the wheel loaders are equipped with seats without suspension in the fore-and-aft direction. The predicting the responses of the combined human body and seat structural responses to the vibration generate by the vehicle is still challenging. This is mainly due to the nonlinearity and complexity of the human body dynamics. Therefore, alternative methods are required to characterize the dynamics of the combined human body–seat structure. A few studies developed methods on the basis of multi-body simulation to reduce the WBV on bigger wheel loaders. This work introduced an automatic simulation process, which was developed based on a full-vehicle model of a wheel loader. With this automatic simulation process, axle and cab suspensions on the wheel loader can be optimized for more riding comfort and operational safety. Other researchers, through help of multi-body models of a wheel loader with suspended axles or unsuspended axles, investigated that WBV in the vertical and the longitudinal directions on the wheel loader was reduced greatly by suspended axles [22]. It's important to highlight that than the education of the drivers has a relevant role in order to prevent the negative effects. The WBV exposure was measured on two different backhoe loaders before and after the operators got an education on eco-driving and vibration avoidance. This training has shown positive results for the purpose of reduce the WBV on backhoe loaders efficiently [23]. The present work is based on a measurement campaign for drivers of haul trucks during a range of normal operating conditions in a basalt quarry site. The aim of the current investigation is to examine the variation of WBV exposures on two different vehicles, one model designed in order to be employed only in the quarry site, the other one, can represent a useful option if employed in different working activities. Once the characteristics of the physical agent have been defined, a safety-oriented organization of working task may be carried out with the focus of reducing WBV exposure by means of different vehicle adoption without determining loss of productivity.

2 MATERIALS AND METHODS

The WBV measurements were taken from two different haul trucks during the LHD activities into a basalt quarry near Rome. The trucks employed in the measurement campaign ranged in capacity from 32 to 80 tonnes. The haul trucks were conducted by two different drivers with similar physical characteristics and they agreed to participate to the tests. Their data are reported in Table 1.

Table 1: Driver's characteristics.

	Age (years)	Weight (kg)	Height (cm)	Driver's experience (years)
Driver 1	34	75	173	10
Driver 2	40	110	185	1

Field tests of vibration measurements were carried out on two different vehicles: Dumper 769 Caterpillar and Mercedes Actros (see Fig. 1). The characteristics of the vehicles are reported in Table 2.

(a)

(b)

Figure 1: (a) 769 Caterpillar model; and (b) Mercedes model.

Table 2: Vehicle's characteristics.

	Weight empty configuration (ton)	Full weight configuration (ton)	Max speed (km)	Year
Dumper 769	45	80	80	2006
Mercedes	15	32	130	2010

Each vehicle is furnished of a seat suspension in the vertical direction comprised of a passive air spring and an inclined hydraulic shock absorber. Instead the driver's seat does not have any suspension in the longitudinal and the lateral directions. The acceleration measurements in the vehicles were realized on the surface and at the base of the drivers' seats in the same roadway conditions inside the quarry site. The test route had a length of 350 m with different sections: for haulage and dump phases were made up of rock smooth surface, with an 8–10° slanting and a rough gravel zone near the load point while (see Fig. 2).The weather was in sunny dry conditions all time long during the summer months and typical temperatures were approximately 27 ± 3°C. The speed of the vehicles ranged from 10–25 km/h, by reason of the automatic transmission system that modified the engine rpm of the truck and consequently the input source signal. WBV exposure were measured according to the standard procedure of the ISO 2631-1 (1997) guidelines.

(a) (b) (c)

Figure 2: (a) Transducer positioning on the driver's seat; (b) Test route inside the quarry site; and (c) Load operation.

A tri-axial seat pad accelerometer (PCB393) was placed on the driver's seat of the truck with regards to direction and anatomical positioning (see Fig. 2). Vibration was measured in a three-axis coordinate system (*x*, *y* and *z*) in order to evaluate the entry points of vibration in worker's body: the *x*-axis to measure vibration in the anterior–posterior direction, the *y*-axis in the medial–lateral direction, and the *z*-axis in the vertical direction. In order to characterize the signal source vibration a uniaxial accelerometer (PCB 393A03) was attached on both vehicles through permanent magnet at the base of the seat. All the accelerometers were connected on three different channels to Soundbook system and the signal was acquired through Samurai software at the sampling frequency of 1250 Hz. The accelerometer was calibrated before each test setting session using PCB calibrator with a signal of 9.8 m/s^2 and 159.2 Hz. Table 3 reports the characteristics.

Table 3: Model, positioning and sensitivity of sampling devices.

Model accelerometer	Position	Sensitivity ($mV/(m/s^2)$)
Tri-axial accelerometer SEN027-PCB	On all vehicles the instrument is fixed on the seat driver in exact point to which corresponds the driver's sacroiliac joints (his/her buttocks)	X axis: 10.52 Y axis: 10.55 Z axis: 10.88
Uniaxial accelerometer PCB 393A03	Axis sensitive in the vertical direction, same direction of the Z axis the tri-axial accelerometer (vertical axis). It was fixed on the chassis on all vehicles at the base of driver's seat.	Z axis: 10.55

In this work, in order to get a representative WBV of the driver for total exposure duration of 8 h (daily vibration exposure), the WBV exposure for each operator was measured for a defined time representative of a standard cycle. The WBV was continuously measured at the driver seat pad under normal conditions for about 2.1 h for Dumper769 truck and 1.6 h for Mercedes model. The typical work cycle of haul trucks is: (1) loading (by a shovel or digger); (2) loaded travel; (3) dumping (unloading); and (4) unloaded travel. The cycle was repeated almost 35 times during the daily shift. The length of each work cycle varied between 10 min and 14 min and was dependent by different factors: (1) the variation of the loading point by the digger machine; (2) the dump zones, near or far by the shatter system; (3) loading waiting time; (4) time waiting during the travel for alternate passage of the trucks in some points due to the narrow road. Measurements obtained on each individual day were taken on haul trucks travelling the same route inside the quarry site. The lap belt was worn in all tests for the safety purpose.

"Dose" of vibration is based on specific relationships between the duration of exposure and vibration magnitude defined in the International Organisation of Standardisation (ISO) standards, allowing the daily vibration exposure to be re-expressed in terms of the equivalent acceleration that would impart the same energy over an 8 h reference period (a notional average working day). This is called the A(8) (m/s^2 r.m.s.). The ISO 2631-1 (1997) guidelines, specifically, indicates the way in order to assess the health risk from the vibration exposure. This guideline is applicable to the vibration in the frequency range from 0.5–80 Hz, transmitted to the body as a whole through the seat pad. The vibration evaluation procedure incorporates a method of averaging vibration level over time and using third octave band frequency spectrum. The RMS acceleration of the frequency weighted vibration (a_w) over a time period provides a measure of vibration level. It is defined by the equation:

$$A(8) = a_w \sqrt{\frac{T_{esp}}{8}} \ \mathrm{m/s^2}, \tag{1}$$

where:
$A(8)$ is the daily dose expressed as (m/s^2);
T_{exp} is the exposure time expressed as (hour);
a_w is the equivalent continuous RMS acceleration also expressed as (m/s^2).

Individual RMS values of the accelerations measured by the precision vibration meter along the x, y, and z directions are represented by a_{wx}, a_{wy}, and a_{wz} respectively. As the damage risk differs along the three axes, a multiplying factor is applied to the frequency weighted vibration values. In the case of WBV, the acceleration values for the two lateral axes (x and y) are multiplied by 1.4, whereas for the vertical axis (z-axis) WBV values are multiplied

$$a_w = \max(K_x * a_{wy}, K_y * a_{wy}, K_z * a_{wz}) \ldots, \tag{2}$$

where

$$K_x = K_y = 1.4; K_z = 1. \tag{3}$$

These K factors are specific weighed coefficient and they are defined in order to consider the non-linearity of human response to different vibration frequencies during the exposure.

(a) (b) (c)

Figure 3: Overview of the LHD cycle. (a) The dumping of the material at the shutter system; (b) The loaded travel; and (c) The unloaded travel.

3 RESULTS

Accelerometer data were collected whilst the haul trucks were involved in normal daily work in a quarry site. The following pictures provides details of measurements realized on each vehicle. The x-axis indicates the time, expressed in second (s) while the y-axis shows the acceleration values (m/s²). Fig. 4 shows time history signal for the 769 model dumper while Fig. 5 shows it for the Mercedes model.

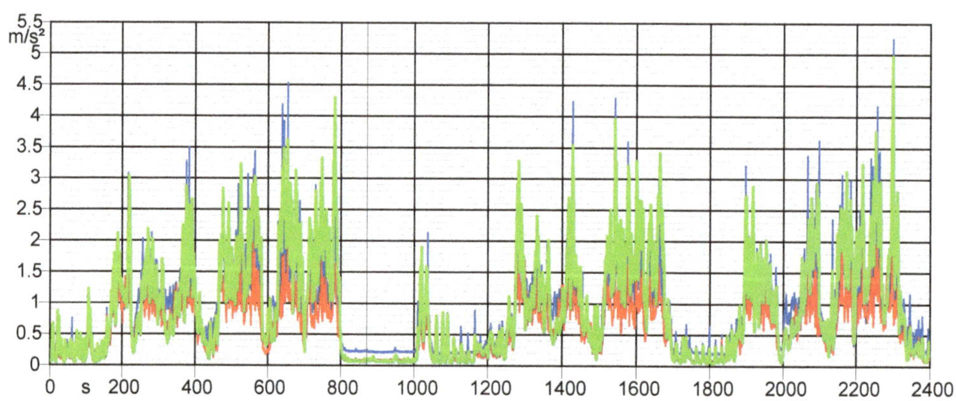

Figure 4: The trend line (blue = x axis; red = y axis; green z = axis) of 769 dumper model.

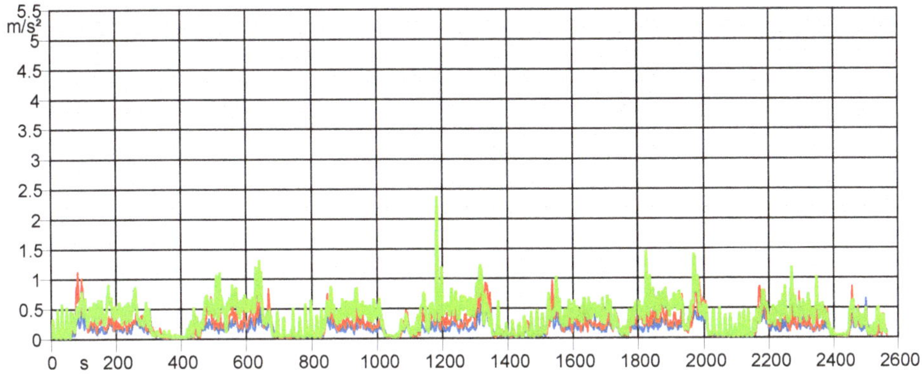

Figure 5: The time history (blue = *x* axis; red = *y* axis; green = *z* axis) of Mercedes model.

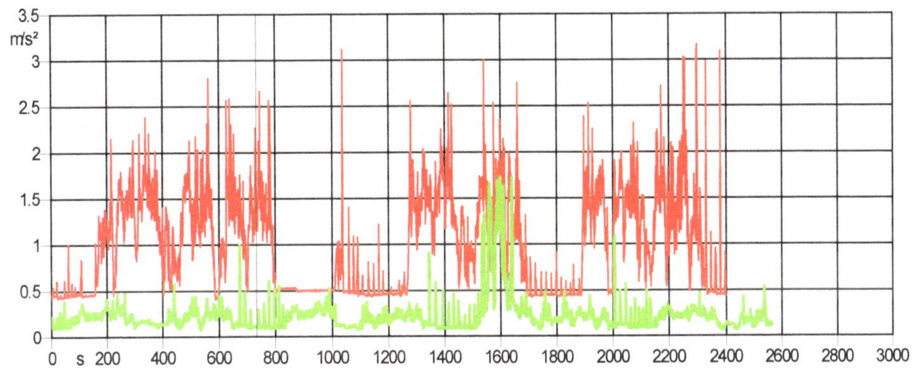

Figure 6: The vibration signal source of the 769 truck (red line) and Mercedes (green line).

The mean frequency-weighted r.m.s WBV of the measurements obtained from the driver's seat of mining haul trucks are presented in Figs 7 and 8. The data are reported in a third band octave. The colours indicates the directions as described above.

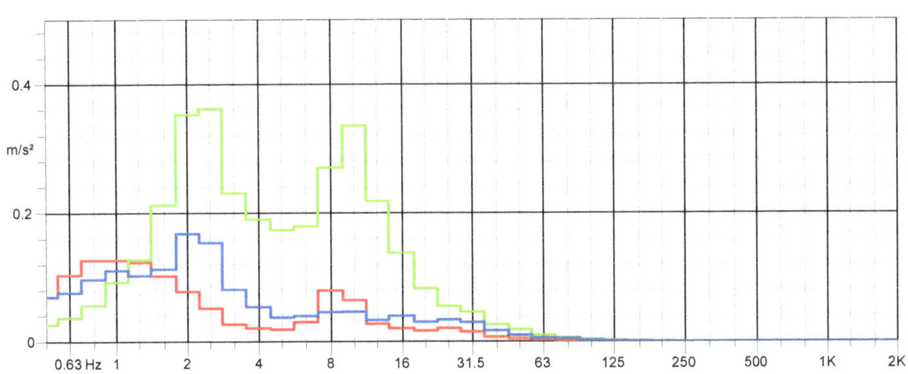

Figure 7: Frequency spectrum of the 769 dumper model.

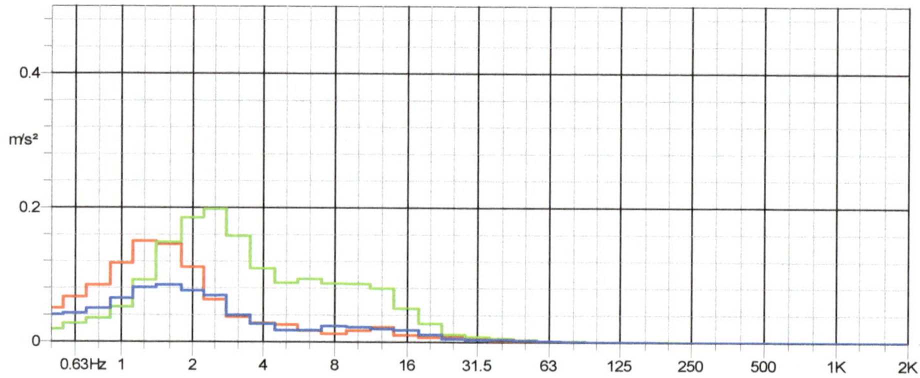

Figure 8: Frequency spectrum of the Mercedes model.

Fig. 9 compares the vibration signal source of the Mercedes vehicles (green line) and the 769 dumper (red line).

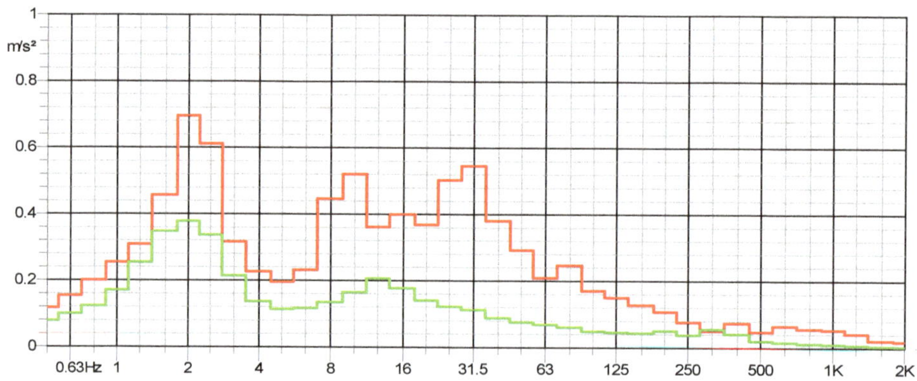

Figure 9: Frequency spectrum of signal source vibration for each vehicle.

As described above on each vehicle near the driver's seat in order a measurement was carried to characterize the signal source vibration. The graph above highlights that the trend of source signal vibration on 769 Caterpillar is similar to that of Mercedes model, with same relevant component at 2 Hz but shows higher values in all frequency spectrum and a relevant frequency component at 31.5 Hz. Table 4 shows the data obtained.

Furthermore, for each measurement, the signal acquired on the driver's seat was analysed in order to evidence the max frequency along direction, the values of parameter $A(8)$ and the peak value. Table 5 sums up sampled values.

Table 4: Comparison of the frequencies and peak values.

Model	$f_{max,z}$ (Hz)	Peak (m/s^2)
Dumper 769	2	3.617
Mercedes	2	1.731

Table 5: Results for both models.

Model	$f_{max,x}$ (Hz)	$f_{max,y}$ (Hz)	$f_{max,z}$ (Hz)	A(8) (m/s^2)	Peak (m/s^2)
Dumper 769	2	1	10	0.498	4.973
Mercedes	1.6	1.3	2.5	0.374	2.361

4 DISCUSSION

A considerable difference in terms of acceleration values was observed in the signal source vibration. The higher vibration intensity is assessed on the Dumper 769 truck. This fact entails that this last model represents a higher vibrational source respect to the Mercedes truck and then the driver of the Dumper 769 suffers a higher daily dose exposure during his activities. Consequently comparing the $A(8)$ values for both vehicles, it is observed that daily exposure estimated according to ISO 2631-1:1997 on Dumper 769 model is much higher than that estimated on the Mercedes model. The $A(8)$ values are based on a 6 hour daily effective exposure. These values are taken into account on the basis of drivers' interview before the starting of measurement campaign. The analysis of sampled data highlights the vertical axis was predominantly associated with the highest frequency weighted accelerations in all measurements, which is consistent with previous investigations of mining haul trucks. The exposure along x and y axes results much lower. The graphs show that y stress direction trend is similar for both trucks. The vibration stress on x and z directions appears always predominant on the 769 Dumper truck. The high acceleration values were recorded in the frequency range 1 and 10 Hz for z axis, in the frequency range from 1–4 Hz along x axis and between 1 and 2 Hz along y axis. The predominant peak frequency weighted accelerations along the z axis was 0.9 m/s^2 at 3.9 Hz. The exposure level doesn't exceed the ISO 2631-1:1997 upper limit on both vehicles, highlighting that the drivers are exposed to caution zone with respect to potential health risks.

5 CONCLUSIONS

The aim of the current investigation is to highlight how different type of vibrational sources may generate a relevant variation of WBV exposure under the same boundary conditions. Nevertheless some common aspects may be underlined the most important of is that all drivers in quarry context are exposed to a very high total vibration magnitude along the vertical z axis. On analysing the outcome of field tests, it may be observed that the same LHD operational cycle may generate a vibrational signal increasing with the weight of the machinery thus determining a difference in terms of daily dose greater than 30% between the two cases. This simple aspect allows some consideration about work organization. In particular quarry management, taking into account these occupational exposure data, is oriented to schedule workers' tasks also aiming to minimize $A(8)$. This target may be addressed by applying a simple FAST approach [24]. This last, well known as Functional Analysis Space (or Machine) Technique, takes into account the machineries necessary for the performance of the single activity together with their a_w value. The formalization of this approach is defined by means of a matrix, which may be stated to represent the space-time discretization of the working task:

	T_1	T_j	T_n
M_1			
M_i		$a_w(i,j)$	
M_m			

The element of the matrix represents the equivalent continuous RMS acceleration of the i-th machine in the interval $T_j = t_{j+1}-t_j$.

The worker's exposure to WBV is instead controlled by the indicating matrix $\{I_k(i,j)\}$, whose element assumes the value of one if the k-th operator utilises the i-th machinery in the time interval $t_{j+1}-t_j$, otherwise it assumes a zero value. The exposure time associated with the k-th operator is therefore:

$$T_E(k) = T \sum_i \sum_j I_k(i,j). \tag{4}$$

Once $T_e(k)$ is assessed it may be inserted in eqn (1) and then the daily exposure is carried out. Such a method, with the outcome of field test allow to minimize exposure data without reducing quarrying productivity.

ACKNOWLEDGEMENTS

The authors wish to acknowledge the fundamental support received from the management, staff and workers of the Area Basalti srl whose full cooperation allowed the development of the research program.

REFERENCES

[1] Rehn, B. et al., Symptoms of musculoskeletal disorders among drivers of all-terrain vehicles in northern Sweden. *NVWW*, **36**, pp. 13–18, 2005.

[2] Orsini, F., Rossi, A., Scorza, A., Botta, F. & Sciuto, S.A., A comparison between a commercial WBV platform and an experimental prototype. *22nd IMEKO TC4 International Symposium and 20th International Workshop on ADC Modelling and Testing 2017: Supporting World Development Through Electrical and Electronic Measurements*, pp. 154–158, 2017.

[3] Cardu, M., Lovera, E. & Patrucco, M., Loading and haulage in quarries: Criteria for the selection of excavator-dumper system. *Proceedings of the 14th International Symposium on Mine Planning and Equipment Selection, MPES 2005 and the 5th International Conference on Computer Applications in the Minerals Industries, CAMI*, 2005.

[4] Monarca, D., Cecchini, M., Guerrieri, M., Santi, M., Bedini, R. & Colantoni, A., Safety and health of workers: Exposure to dust, noise and vibrations. *VII International Congress on Hazelnut*, **845**, pp. 437–442, 2008.

[5] Gerard, F. & Pierre, M., Numerical assessment of fore-and-aft suspension performance to reduce whole-body vibration of wheel loader drivers. *Journal of Sound and Vibration*, **298**, pp. 672–687, 2006.

[6] Alfaro Degan, G., Coltrinari, G., Lippiello, D. & Pinzari, M., A comparison between methods for assessment of whole-body vibration exposure: a case study in a limestone quarry. *International Journal of Safety and Security Engineering*, **8**(1), pp. 90–97, 2018.

[7] Bovenzi, M. & Hulshof, C.T.J., An updated review of epidemiologic studies on the relationship between exposure to whole-body vibration and low back pain. *J. Sound Vib.*, **215**, pp. 595–611, 1998.

[8] Atkinson, S., Robb, M., Mansfiel, D. & Neil, J., Long-term vibration dose for truck drivers: Preliminary results and methodological challenges. *Proceedings of the 37th United Kingdom Conference on Human Responses to Vibration*, 2002.

[9] Paddan, G.S., Mansfield, N.J., Arrowsmith, C.I., Rimell, A.N., King, S.K. & Holmes, S.R., The influence of seat backrest angle on perceived discomfort during exposure to vertical whole-body vibration. *Ergonomics*, **55**(8), pp. 923–936, 2012.

[10] Kumar, S., Vibration in operating heavy haul trucks in overburden mining. *Applied Ergonomics*, **35**(6), pp. 509–520, 2004.

[11] Bureau of Labor Statistics, *Nonfatal Occupational Injuries and Illnesses Requiring Days Away from Work 2014*, Bureau of Labor: Statistics, US Department of Labor, 2015.

[12] Scarlett, A.J. & Stayner, R.M., Whole-body vibration on construction, mining and quarrying machines: Evaluation of emission and estimated exposure levels. HSE Report, 400, 2005.

[13] Alem, N., Application of the new ISO 2631-5 to health hazard assessment of repeated shocks in US army vehicles. *Industrial Health*, **43**(3), pp. 403–412, 2005.

[14] Burgess-Limerick, R., How on earth moving equipment can ISO-2631 be used to evaluate WBV exposure? *J. Safety Health Res. Pract.*, **4**(2), pp. 13–21, 2012.

[15] Wolfgang, R. & Burgess-Limerick, R., Whole-body vibration exposure of haul truck drivers at a surface coal mine. *Applied Ergonomics*, **45**(6), pp. 1700–1704, 2014.

[16] Zhao, X. & Schindler, C., Evaluation of whole-body vibration exposure experienced by operators of a compact wheel loader according to ISO 2631-1:1997 and ISO 2631-5:2004. *International Journal of Industrial Ergonomics*, **44**(6), pp. 840–850, 2014.

[17] Smets, M.P.H., Eger, T.R. & Greiner, S.G., Whole-body vibration experienced by haulage truck operators in surface mining operations: A comparison of various analysis methods utilized in the prediction of health risks. *Appl. Ergon.*, **41**(6), pp. 763–770, 2010.

[18] Chaudhary, D.K., Bhattacherjee, A., Patra, A.K. & Chau, N., Whole-body vibration exposure of drill operators in iron ore mines and role of machine-related, individual, and rock-related factors. *Safety and Health at Work*, **6**(4), pp. 268–278, 2015.

[19] Eger, T., Stevenson, J., Boileau, P.E., Salmoni, A. & Vi, R.G., Predictions of health risks associated with the operation of load–haul–dump mining vehicles: Part 1 Analysis of whole-body vibration exposure using ISO 2631-1 and ISO-2631-5 standards. *Int. J. Ind. Ergon.*, **38**, pp. 726–738, 2008.

[20] Paddan, G.S. & Griffin, M.J., Use of seating to control exposures to whole-body vibration. *Health and Safety Executive*, 2001.

[21] Yuming, Y., Subhash, R., Jue, Y. & Boileau, P., Characterization of a hydro-pneumatic suspension strut with gas-oil emulsion. *Mechanical Systems and Signal Processing*, **106**, pp. 319–333, 2018.

[22] Rehnberg, A. & Drugge, L., Ride comfort simulation of a wheel loader with suspended axles. *Int. J. Vehicle Syst. Model. Test.*, **3**, pp. 168–188, 2008.

[23] Langer, T.H., Iversen, T.K., Andersen, N.K., Mouritsen, O.Ø. & Hansen, M.R., Reducing whole-body vibration exposure in backhoe loaders by education of operators. *Int. J. Ind. Ergon.*, **42**, pp. 304–311, 2012.

[24] Alfaro Degan, G., Lippiello, D. & Pinzari, M., A geostatistical approach to the functional analysis space technique: A case of study. *Advances in Safety and Reliability: Proceedings of 16th European Safety and Reliability Conference (ESREL)*, **1**, pp. 45–52, 2005.

COMPUTER-AIDED HAZOP STUDIES: KNOWLEDGE REPRESENTATION AND ALGORITHMIC HAZARD IDENTIFICATION

JOHANNES SINGLE, JÜRGEN SCHMIDT & JENS DENECKE
CSE Center of Safety Excellence (CSE-Institut), Germany

ABSTRACT

Safety assessments are conducted to identify and assess the risks that arise from processes, process plants or technical systems in general. This includes the identification of potential hazards posed by plants. One recognized and generally accepted method for this is the hazard and operability (HAZOP) method. It is a human-centered process that is time- and labor-intensive. In the presented research approach, the structure of a computer-aided HAZOP system is described. The identification of hazards and malfunctions within technical systems is knowledge-intensive. Within this research approach, it transpired that the semantically correct and detailed modeling of deviation cause and effect relationships in the form of ontologies are of particular importance to draw correct conclusions. Thus, the guiding principles of a knowledge representation framework are described from a process safety perspective, and serve as a basis for the automatic identification of hazards. An integral understanding of the process, process plant and involved substances requires extensive knowledge. The way in which this knowledge is used and the search for hazards is conducted has an influence on the completeness of the results. Within this approach, the hazard/malfunction identification is conducted on different layers of abstraction to improve the efficiency of the search algorithm. The proposed methodology is applied within a case study to a technical system that consists of a compressor, vessel and valve. The first results demonstrate that the proposed method is well-suited to understand and identify the context of hazards and malfunctions. Thus, a system for computer-aided HAZOP studies can be used to assist HAZOP conductors in performing hazard analysis while increasing the speed of safety assessments and serving as a decision support system.

Keywords: computer-aided HAZOP study, hazard identification, ontology-based reasoning, decision support system, automatic hazard analysis.

1 INTRODUCTION

Process hazard analysis (PHA) methods are used to identify and evaluate hazardous events that can occur in process plants. Within the chemical process industry, the HAZOP study is an established technique to review the process plant and identify potential hazardous events and operability malfunctions (see Kletz [1]). In HAZOP studies, the process plant is systematically and critically analyzed. It is thereby divided into sections called nodes, which are easier to analyze compared to the entire process plant. This is usually done based on the plant's Piping and Instrumentation Diagrams (P&ID). Afterwards, process deviations are applied to the nodes, while possible causes, consequences and safeguards are identified and discussed. Usually, HAZOP studies are conducted by a team of experts from different fields, such as plant operation, process control and safety engineering. The HAZOP study is a human-centered, guided brainstorming technique. Thus, it is a time- and labor-intensive process that is not always carried out for the entire process plant, but for particularly critical sections. The results of the study depend on the knowledge, personal experience, moderation, communication, discussion culture and level of training of the HAZOP study practitioners. Furthermore, the capabilities of the practitioners can be limited by repetitive tasks, large amounts of data or stressful conditions. Hence, human experts can be supported by using computer-aided HAZOP (CAH) systems, which represent computer programs that serve as

WIT Transactions on The Built Environment, Vol 189, © 2019 WIT Press
www.witpress.com, ISSN 1743-3509 (on-line)
doi:10.2495/SAFE190061

Decision Support Systems. Such systems can be used to remove repetitive tasks and analyze large sets of data in HAZOP studies. Also, plant operators can be supported in the early stages of the HAZOP process, in which a safety professional may not necessarily be present. The usability of CAH systems depends on the quality of their hazard identification abilities and the completeness and reliability of the results.

1.1 Brief state of research

The automation of elements of HAZOP studies has been a topic of research for more than 30 years. The first systems were mainly built on logic trees, rules and rule networks, within expert system shells (see [2]–[4]). Later research approaches used qualitative models for reasoning purposes in combination with rules (see [5]–[8]). Eizenberg et al. [9] used quantitative model-based reasoning approaches based on process simulations. Some approaches aimed to integrate computer-aided design (CAD) software with automatic hazard identification. More recent developments used sophisticated graph-based methods (see [10]–[12]). Zhao et al. [13] used case-based reasoning to draw conclusions based on previous situations.

1.2 Technical basis

A computer program for computer-aided HAZOP studies is essentially concerned with three tasks: (1) representation of the process plant, (2) knowledge representation of generic and process-specific knowledge, and (3) hazard or malfunction identification. The available technologies with which to achieve these tasks are briefly described below.

The technical system or process plant must be digitally represented in order to be analyzed. Thus, information regarding the arrangement of plant units and interconnections is needed. This can be generated using a graphical editor, although this can be time-consuming for complex systems. Another possibility is the direct extraction of plant information from a P&ID or Process Flow Diagram (PFD). This information can also be extracted using process simulation programs. This automatic form of extraction has the advantage of reducing the time that is required to model the process plant. In order to reuse process plant information, a shared data model is crucial to reduce the data translation effort. For instance, ISO 18629-1:2004 or DIN EN 62424:2017-12 describe exemplary data models (see [14], [15]).

The term "knowledge" represents the understanding of information, facts or skills. Knowledge must be made available to human experts and also to machines. Hence, a knowledge representation formalism is needed, which has different requirements:

- Various types of knowledge, such as facts and procedures, must be recorded;
- The semantic context of knowledge fragments must be clearly illustrated;
- Knowledge structures must be easily extensible; and
- Knowledge queries and inference must be efficient and complete.

There are various formalisms for knowledge representation, such as rules, semantic-nets in [16], frames in [17] and ontologies in [18]. Rules are formulated to model relationships. The semantic context of strict rules is difficult to map, so there must be a rule for every single relationship, while the context cannot be illustrated. In semantic-nets, knowledge is stored using graphs, where nodes represent objects while the edges (connections) represent the relationships between the objects. Frames were derived from semantic-nets and can be used to represent stereotypical situations, but frames lack formal semantics (see van Harmelen et al. [19]), which means that there is no explicit notation to illustrate the meaning of the knowledge. This can lead to ambiguity. Furthermore, there are ontologies, which are formal definitions of types, properties and relationships of knowledge fragments. Due to their formal semantics, ontologies can be used to unambiguously represent knowledge. The usage of ontologies for HAZOP studies has already been proposed by Batres [20] and Daramola et al. [21]. Other research groups have used ontologies for decision support in [22] and process supervision in [23].

The identification of hazards is conducted based on knowledge models and a digital representation of the process plant. Thus, facts and relationships must be extracted from these knowledge models, and can be achieved using various reasoning techniques. Here, "reasoning" describes the process of drawing conclusions and making inferences. There are various reasoning technique classes that can be used for hazard identification, such as rule-based, case-based, and model-based reasoning (see [24]–[26]). Rule-based reasoners attempt to match rules from a new case with rules from a knowledge base. In case-based reasoning, a previous situation that is similar to the current one is used to solve the current problem. Model-based reasoners use models to draw inferences. These models represent the behavior or structure of a system, and can be quantitative (based on equations) or qualitative (based on a behavioral description). The type of knowledge representation has a direct effect on the reasoning technique and vice versa. For instance, case-based reasoning cannot be directly applied to a knowledge representation that is well-suited to rule-based reasoning.

2 OBJECTIVES

The aim of this research project is to identify possible applications for computer-aided HAZOP systems and develop a comprehensive decision support system. The automatic hazard identification within the decision support system is based on complex knowledge models in the form of ontologies. In addition, different methods for the identification of hazards must be identified and tested, while methods that use artificial intelligence are analyzed.

In this paper, the focus is on the representation of knowledge using knowledge models and algorithmic hazard identification. There are sufficient technologies and data model structures to reuse digital process plant information for hazard identification purposes. Our own approach includes multilayer automatic hazard identification, which is based on an appropriate knowledge representation from a process safety perspective. Hence, the representation of knowledge is considered in particular detail. The following aims are addressed:

- Propose required knowledge domains from a process safety perspective;
- Present a knowledge structure that is suited for computer-aided HAZOP systems;
- Determine and establish different abstraction layers for the identification of hazards;
- Present a conceptual framework for a hazard identification algorithm;
- Demonstrate the methodology within a case study of a technical system; and
- Present the limitations and areas in which further research is needed.

3 METHODOLOGY

In the following section, the knowledge models (3.1), hazard identification on different layers (3.2), ontology implementation (3.3), guiding principles behind the ontologies (3.4) and hazard identification algorithm (3.5) are described.

3.1 Knowledge representation through knowledge models

In order to supply a computer-aided HAZOP system with knowledge, the knowledge representation must contain knowledge from various domains, such as

- Information regarding substances, e.g., hazardous properties or chemical reactivity;
- (Chemical) processes and process-related procedures;
- Process engineering knowledge to support a context-aware structure;
- HAZOP-specific knowledge, e.g., guidewords and the effect of deviations on processes;
- Knowledge regarding process units, equipment, subsystems and environment;
- Typical and atypical failure modes of plant components;
- Potential causes and consequences of failure modes;
- Safeguards to prevent hazardous events to mitigate failures or malfunctions; and
- Information about former events and incidents.

The required knowledge needs to be structured and formulated explicitly, while the semantic contexts must be well-defined. This can be achieved using ontologies for knowledge representation. Within this work, the knowledge representation framework is divided into *conceptual* and *application-oriented* ontologies. Ontologies can be used to make definitions coherent, organize knowledge clearly and unambiguously describe concepts. Hence, they are well-suited to describe relations between concepts and can be formalized using formal semantics. Furthermore, they are machine manipulable, which means that they are machine readable and writable. Currently, the designed ontologies cover the following domains:

- Process engineering principles, e.g., processes and relationships between process-variables;
- HAZOP, e.g., guidewords, deviations, and the propagation of deviations;
- Cause and consequence, e.g., deviation–cause–consequence relationships;
- Material, e.g., hazardous properties and chemical interactions; and
- Plant, e.g., the hierarchical structure of units, equipment or plant components.

3.2 Hazard identification on different layers of abstraction

Within this work, the identification of hazards is based on the knowledge models and a representation of the process plant. With regard to hazard identification, the integral consideration of the plant transpired to be too complex for an algorithm to analyze. This was due to the fact that an integral understanding of the process, the process plant and the involved substances and the interactions between substances requires extensive knowledge. The way in which this knowledge is used for hazard identification is important in order to draw correct conclusions. Thus, the introduction of layers of abstraction makes a search algorithm more efficient, since the consideration of additional information restricts the search range. The identification of hazardous events and operational malfunctions is conducted on four layers in this work: (1) Substance Layer, (2) Specific Unit Layer, (3) Abstract Object Layer and (4) Fault/Hazard Propagation Layer. These layers are illustrated in Fig. 1.

The hazard and malfunction identification at different levels of abstraction brings advantages, such as different degrees of detail, the avoidance of redundancies, the propagation of hazards and malfunctions, and the propagation of cause–effect relationships through the process plant. Furthermore, not all layers are needed to produce a rough estimation of hazards. The layers are described in greater detail below:

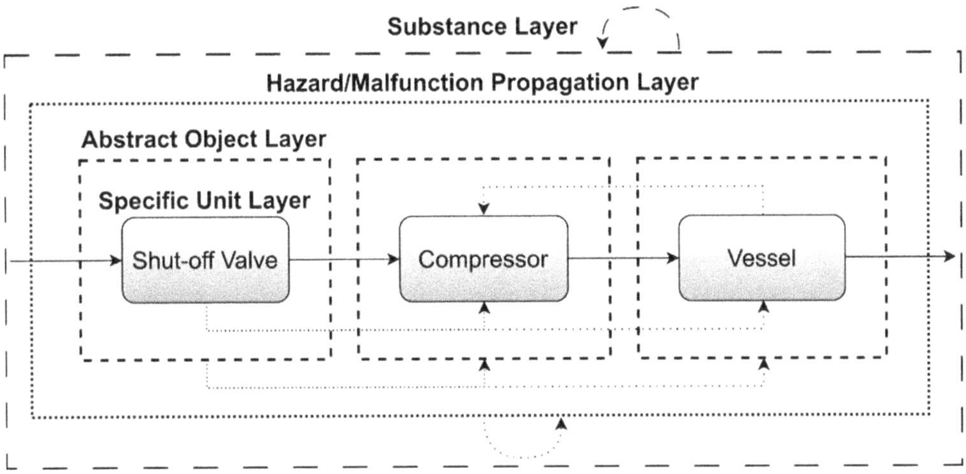

Figure 1: Schematic representation of the different layers of abstraction.

1. Substance Layer: the hazardous properties of substances, e.g., flammable;
2. Specific Unit Layer: equipment-specific hazards and malfunctions, e.g., faulty compressor speed;
3. Abstract Object Layer: direct periphery of equipment (e.g., oil-cooler), operating condition (e.g., start-up operation), type of equipment (e.g., machine), component of equipment (e.g., bearings or seals), and function (e.g., conveying of gases); and
4. Fault/Hazard Propagation Layer: interactions between equipment and the propagation of cause and effect, e.g., more flow + confinement = more pressure.

3.3 Ontology implementation

There are various ontological languages, such as Resource Description Framework (RDF) language and the Web Ontology Language (OWL). The RDF language is used to describe structured information within simple ontologies. OWL represents a compromise between expressiveness and efficient reasoning and is the standard modeling language for ontologies as recommended by the World Wide Web Consortium (W3C) in 2004 (see [27]). Furthermore, an improved version of the language, OWL2, has been produced. The basic elements of the OWL language are *classes* (concepts), *properties* (relations) and *individuals* (instances), while class relationships and restrictions are used to increase the semantic clarity.

The OWL2 ontologies in this work were designed and built using Owlready2 (see Lamy [28]), which is a module for ontology-oriented programming in Python. The HermiT reasoner (see Glimm et al. [29]) was included in Owlready2 and used for inferring logical consequences from the asserted facts and axioms within the ontologies. In this case, ontology-based reasoning was based on the inference rules that were specified by the ontology language, i. e., OWL2.

3.4 Guiding principles behind the designed ontologies

In this work, guiding principles for the modeling of deviation–cause–consequence relationships are developed and used to model relationships in a semantically correct and consistent way. These guiding principles are explained in detail below (see eqns (1)–(5)).

Deviations from the target state of process parameters are described qualitatively. The class *Deviation* is composed of the classes *Parameter* and *Guideword*:

$$Parameter + Guideword \ \Rightarrow Deviation. \tag{1}$$

An exemplary relation of eqn (1) is presented in Table 1 using the Owlready2 (Python) syntax. For instance, the *Deviation* class *HighTemperature* is composed of the classes *More* and *Temperature* and the properties *hasGuideword* and *hasParameters.*

Table 1: Exemplary Owlready2 syntax of the *HighTemperature* class.

```
class HighTemperature(Deviation):
    equivalent_to = [Deviation &
                     hasGuideword.some(More) &
                     hasParameter.some(Temperature)]
```

According to the developed knowledge model concept, *Causes* can be constructed from the *Deviation* and the *Abstraction level*:

$$Deviation + Abstraction\ level \ \Rightarrow Cause. \tag{2}$$

The *Abstraction level* includes specific details about the contexts of the causes, such as the subsystem of the equipment or components and the function of the (process) equipment. Furthermore, an abstraction level represents additional information, such as hazardous properties or the state of aggregation. The semantic context of this is illustrated in Fig. 2. In addition, an exemplary cause is shown in Table 2, where the cause *ExcessiveDischargeTemperature* is constructed from the deviation *HighTemperature* and the properties *takesPlaceIn* and *hasStateOfAggregation.* The semantic context of these properties is also presented in Fig. 2.

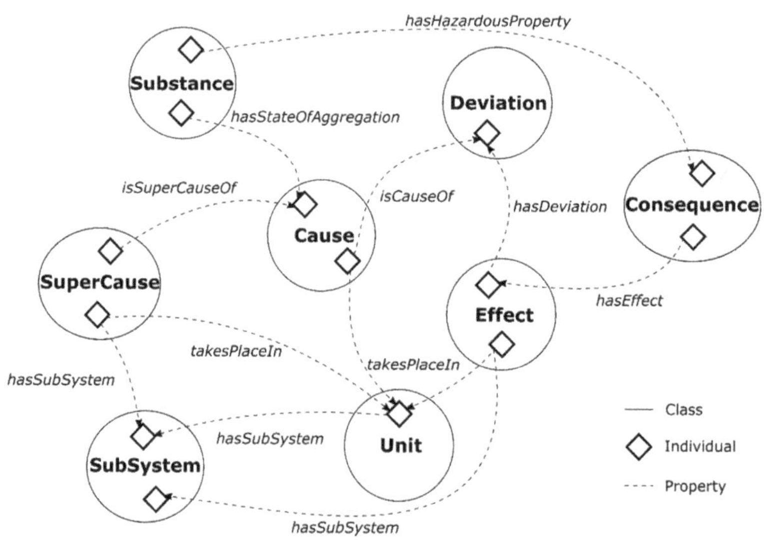

Figure 2: Visual representation of the concept graph of classes, its individuals and its semantic connection through properties as part of a process safety knowledge base.

Table 2: Exemplary Owlready2 syntax of the *ExcessiveDischargeTemperature* cause class.

```
class ExcessiveDischargeTemperature(Cause):
    equivalent_to = [Cause &
                     hasStateOfAggregation.some(Gaseous) &
                     takesPlaceIn.some(Compressor) &
                     isCauseOf.some(HighTemperature)]
```

Causes can have a superordinate cause, which is defined as a *SuperCause* (see Fig. 2):

$$Cause + Abstraction\ level\ \Rightarrow SuperCause. \tag{3}$$

For instance, the cause *ExcessiveDischargeTemperature* could have the superordinate cause *FaultyLubrication*. This distinction enables a more detailed representation of causal relationships.

Deviations can also lead to consequences such as hazardous events or malfunctions. In this approach, (*intermediate*) *Effects* and *Consequences* are distinguished. Within our model, representation consequences follow effects, e.g., the effect is the leakage of a flammable liquid, and the consequence is ignition and fire. Furthermore, *Effect* knowledge models are composed from the deviations and abstraction level details:

$$Deviation + Abstraction\ level \Rightarrow Effect. \tag{4}$$

The exemplary effect model of a *Rupture* is presented in Table 3. In this model, concept effects lead to consequences. For instance, an effect could be a *Rupture*, for which the corresponding consequence could be a *LossOfContainment*. A Consequence (see Fig. 2) is derived from the Effects and the details from the Abstraction level:

$$Effect + Abstraction\ level \Rightarrow Consequence. \tag{5}$$

For instance, the consequence *ToxicRelease* has the effect *LossOfContainment* and the hazardous property *Toxic* within the abstraction level.

Table 3: Exemplary Owlready2 syntax of the Rupture effect class.

```
class Rupture(Effect):
    equivalent_to = [Effect &
                     hasStateOfAggregation.some(Gaseous | Liquid) &
                     (hasDeviation.some(HighTemperature) &
                     takesPlaceIn.some(PressurizedContainment | Piping)) |
                     hasDeviation.some(HighPressure)]
```

3.5 Methodology and hazard identification algorithm

A program flow chart of the proposed method is provided in Fig. 3. First, HAZOP deviations are generated. Then, the first unit is selected and all possible deviations are applied to it. For every deviation possible, (super) causes, effects and consequences are inferred. Thus, all ontologies involved are evaluated simultaneously on abstraction layers 1–3 (see Section 3.4). In case there are more units, the procedure is repeated while the propagation of deviations, hazards or malfunctions between units is considered on the abstraction layer (4).

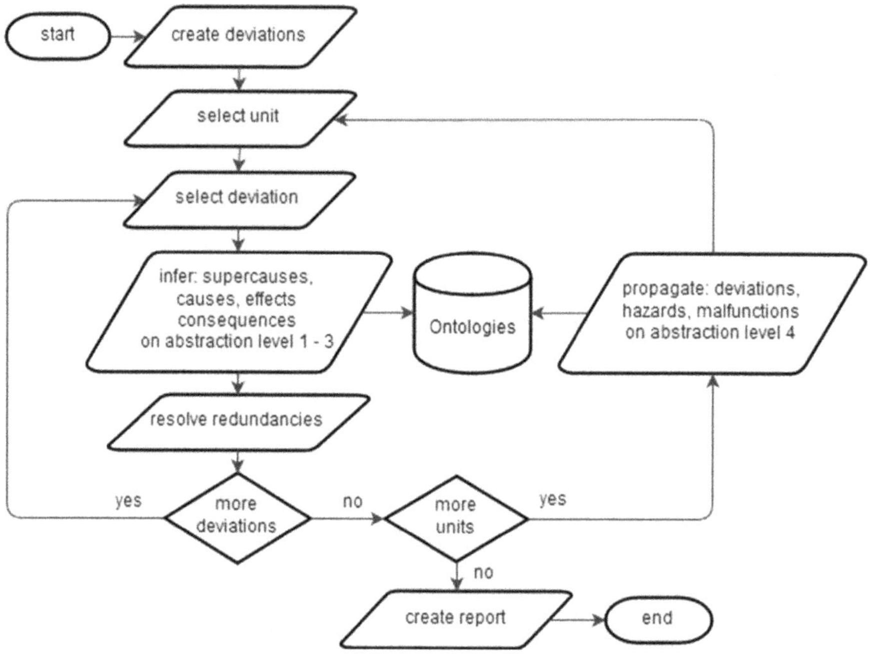

Figure 3: Flow chart of the method for implementing the hazard identification algorithm.

4 CASE STUDY

The methodology presented was applied to a process plant within a case study. Therefore, a comparatively simple system without chemical reactions or a complex arrangement of components was chosen to illustrate the concept and the generated results. This system consists of a compressor, a vessel, a shut-off valve and connection pipes (see Fig. 4) and is operated using air. The compressor takes in (dry) air at ambient pressure (1 bar), compresses it (15 bar) and conveys it into the vessel, which is shut off by a valve. Furthermore, the compressor has an oil-cooling subsystem. Hence, besides air, oil is present in the piston compressor. The shut-off valve is pneumatically actuated.

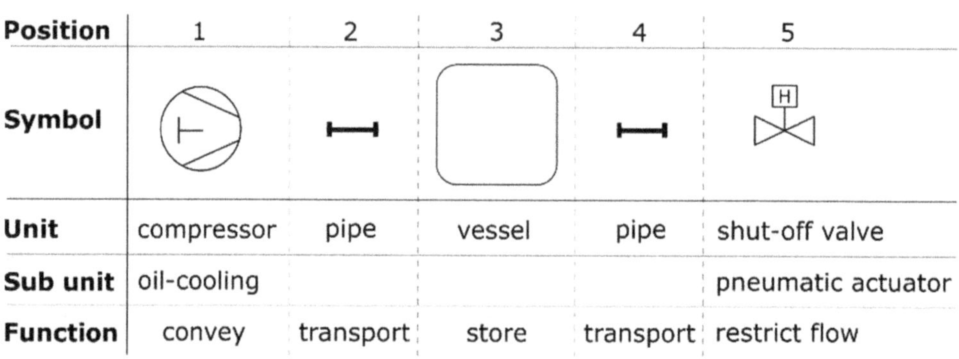

Figure 4: Exemplary compressor-vessel system.

The procedure of the algorithm is explained on the basis of Fig. 3. First, the deviations (see Tables 4 and 5) are generated from the process parameters "flow" and "pressure" and the guidewords "no", "less", and "more". These process parameters and guidewords have only been selected as an example. After the deviations have been generated, the first unit (compressor) is selected and the deviations are applied to find (super)causes, effects and consequences on the *Specific unit layer* and *Abstract object layers.* For instance, the supercause *LiquidHammer* is detected on the *Specific unit layer.* Additionally, an investigation on the *Substance layer* shows that oil has the "flammable" hazardous attribute. Taking the *Abstract object layer* into account, a defective seal (cause) can lead to leakage (effect) and fire (consequence). This coherence is not bound to a specific deviation (see Table 4). Subsequently, the next deviation is applied and the procedure is repeated (see Fig. 3). After the examination of the first unit, the next unit is examined. This procedure is continued until no units remain.

Table 4: Exemplary automatically generated results of the compressor (incomplete).

Plant component: **Piston compressor**

Deviation	SuperCause	Cause	Effect	Consequence
noFlow	Insufficient-Lubrication; MalfunctionCooling PowerFailure; LiquidSlugging	Failure	Faulty-Plant	LossOf-Production
highFlow	MalfunctionSpeed-Control; IncreasedSuction-Pressure	DecreasedDifferential-Pressure	[*Propagation*]	None
highTemperature	MalfunctionCooling-System; InsufficientVentilation	ExcessiveDischarge-Temperature	[*Propagation*]	None
–	Embrittlement	DefectiveSeal	Leakage (oil)	Fire

Table 5: Exemplary automatically generated results of the vessel (incomplete).

Plant component: **Vessel**

Deviation	SuperCause	Cause	Effect	Consequence
highPressure	AbnormalHeatInput	ThermalExpansion; AbnormalGasIntake	Rupture	Blast
lowPressure	Corrosion; Erosion; MechanicalDamage; DefectiveSeal	Leakage	LossOf-Containment	None
lowTemperature	FastGasRelaxation	AbruptReliefOf-VesselContent	BrittleFracture	None

From the second unit onward, the *Hazard propagation layer* is considered (see Fig. 3). For instance, valves or pressure raisers are recognized because they can have a direct effect on upstream or downstream equipment.

The consideration of the *Hazard propagation layer* (marked by [Propagation]) shows that the deviation *highFlow* in the compressor and a closed shut-off valve lead to the deviation *highPressure* in the vessel. This connection is not apparent when looking at the individual units (see Tables 4 and 5). The adjustment of the *Specific unit layer* and *Abstract object layers* prevents redundant results and also ensures plausibility, since the context of hazards and malfunctions are given greater consideration on the *Abstract object layer*. For instance, the subsystem oil-cooling is considered on the *Abstract object layer* which led to the potential consequence *Fire* (see Table 4).

5 CONCLUSIONS AND FUTURE WORK

The automation of HAZOP studies, or more precisely, the automation of selected components of HAZOP studies (such as the identification of hazards), has long been the topic of research for more than 30 years. The identification of hazards is a knowledge-intensive process and the quality and semantic richness of knowledge models has a direct effect on the quality and completeness of a safety assessment. In this paper, a knowledge structure for computer-aided HAZOP systems has been proposed. Thus, coherent, context-aware and semantically rich knowledge models in the form of ontologies were designed, formalized and implemented in this research approach, from a process safety perspective. The model was implemented using the OWL ontology language within Owlready2. The semantically correct and complete modeling of deviation–cause–consequence relationships transpired to be particularly important. The hazard identification within this approach was conducted based on the proposed knowledge structure on different layers of abstraction: (1) Substance Layer, (2) Specific Unit Layer, (3) Abstract Object Layer, and (4) Hazard/Malfunction Propagation Layer. The results demonstrate that the investigation on different layers led to more complete and fewer redundant results. Furthermore, the proposed methodology and its principles are not solely limited to process plants and can be applied to other technical systems. Therefore, the ontologies must be extended and refined with further knowledge. Also, case studies of more complex systems, such as chemical reactors, must be conducted in order to validate the proposed methodology. Moreover, the ontologies can be used to model accidents or hazardous incidents and can therefore serve as a coherent long-term digital memory and a comprehensive process safety knowledge base. The focus of future research should be the consideration of chemical reactions and safeguards, the usage of an existing upper ontology, and the enhancement of the hazard identification algorithm using case-based reasoning.

REFERENCES

[1] Kletz, T.A., Hazop: Past and future. *Reliability Engineering and System Safety*, **55**, pp. 263–266, 1997.

[2] Shum, S.K., Davis, J.F., Punch, W.F. & Chandrasekaran, B., An expert system approach to malfunction diagnosis in chemical plants. *Computers and Chemical Engineering,* **12**, pp. 27–36, 1988.

[3] Bassiliades, N., Governatori, G. & Paschke, A., Rule-based reasoning, programming, and applications. *Proceedings of the 5th International Symposium, RuleML*, pp. 305–312, 2011.

[4] Khan, F.I. & Abbasi, S., Tophazop: A knowledge-based software tool for conducting hazop in a rapid, efficient yet inexpensive manner. *Journal of Loss Prevention in the Process Industries*, **10**, pp. 33–343, 1997.

[5] Suh, J.C., Lee, S. & Yoon, E.S., New strategy for automated hazard analysis of chemical plants. Part 1: Knowledge modelling. *Journal of Loss Prevention in the Process Industries,* **10**, pp. 113–126, 1997.

[6] Kang, B., Lee, B., Kang, K., Suh, J. & Yoon, E., Aha: A knowledge based system for automatic hazard identification in chemical plant by multimodel approach. *Expert Systems with Applications*, **16**, pp. 183–195, 1999.
[7] Vaidhyanathan, R. & Venkatasubramanian, V., Experience with an expert system for automated hazop analysis. *Computers and Chemical Engineering*, **20**, pp. 1589–1594, 1996.
[8] McCoy, S.A., Wakeman, S.J., Larkin, F.D., Chung, P., Rushton, A.G. & Lees, F.P., Hazid: A computer aid for hazard identification. *Process Safety and Environmental Protection*, **77**, pp. 328–334, 1999.
[9] Eizenberg, S., Shacham, M. & Brauner, N., Combining HAZOP with dynamic simulation: Applications for safety education. *Journal of Loss Prevention in the Process Industries*, **19**, pp. 754–761, 2006.
[10] Batres, R., Fujihara, S., Shimada, Y. & Fuchino, T., The use of ontologies for enhancing the use of accident information. *Process Safety and Environmental Protection*, **92**, pp. 119–130, 2014.
[11] Rossing, N.L., Lind, M., Jensen, N. & Jørgensen, S.B., A functional hazop methodology. *Computers and Chemical Engineering*, **34**, pp. 244–253, 2010.
[12] Rodríguez, M. & de La Mata, J.L., Automating hazop studies using d-higraphs. *Computers and Chemical Engineering*, **45**, pp. 102–113, 2012.
[13] Zhao, J., Cui, L., Zhao, L., Qiu, T. & Chen, B., Learning hazop expert system by case-based reasoning and ontology. *Computers and Chemical Engineering*, **33**, 371–378, 2009.
[14] ISO 18629-1:2004, Industrial automation systems and integration: Process specification language – Part 1: Overview and basic principles. International Organization for Standardization, 2004.
[15] DIN EN 62424:2017-12, Representation of process control engineering: Requests in P&I diagrams and data exchange between P&ID tools and PCE-CAE tools (IEC 62424:2016). German Institute for Standardisation (DIN, Deutsches Institut fr Normung e.V.), 2017.
[16] Studer, R., Grimm, S. & Abecker, A. (eds), *Semantic Web Services*, Springer: Berlin and Heidelberg, pp. 51–105, 2007.
[17] Minsky, M., A framework for representing knowledge. *Computation and Intelligence,* ed. G.F. Luger, American Association for Artificial Intelligence: Menlo Park, CA, pp. 163–189, 1995.
[18] Bittner, T., From top-level to domain ontologies: Ecosystem classifications as a case study. *Proceedings of the 8th International Spatial Information Theory Conference*, pp. 61–77, 2007.
[19] van Harmelen, F., Lifschitz, V. & Porter, B. (eds), *Handbook of Knowledge Representation*, Elsevier: Amsterdam, pp. 135–179, 2008.
[20] Batres, R., An ontology approach to support HAZOP studies. *Proceedings of the Asian Pacific Confederation of Chemical Engineering Congress Program and Abstract,* 2004.
[21] Daramola, O., Stålhane, T., Omoronyia, I. & Sindre, G., Using ontologies and machine learning for hazard identification and safety analysis. *Managing Requirements Knowledge,* eds W. Maalej & A.K. Thurimella, Springer: Heidelberg, pp. 117–141, 2013.
[22] Baumeister, J. & Striffler, A., Knowledge-driven systems for episodic decision support. *Knowledge-Based Systems*, **88**, pp. 45–56, 2015.

[23] Musulin, E., Roda, F. & Basualdo, M., A knowledge-driven approach for process supervision in chemical plants. *Computers and Chemical Engineering*, **59**, pp. 164–177, 2013.

[24] Dutta, S. & Bonissone, P.P., Integrating case- and rule-based reasoning. *International Journal of Approximate Reasoning*, **8**, pp. 163–203, 1993.

[25] Kolodner, J.L., An introduction to case-based reasoning. *Artificial Intelligence Review*, **6**, pp. 3–34, 1992.

[26] Khardon, R. & Roth, D., Defaults and relevance in model-based reasoning. *Artificial Intelligence*, **97**, pp. 169–193, 1997.

[27] Hitzler, P., Krotzsch, M. & Rudolph, S., *Foundations of Semantic Web Technologies*, CRC Press: New York, pp. 111–210, 2010.

[28] Lamy J.B., Owlready: Ontology-oriented programming in Python with automatic classification and high level constructs for biomedical ontologies. *Artificial Intelligence In Medicine*, **80**, pp. 11–28, 2017.

[29] Glimm, B., Horrocks, I., Motik, B., Stoilos, G. & Wang, Z., HermiT: An OWL 2 reasoner. *Journal of Automated Reasoning*, **53**(3), pp. 245–269, 2014.

DESIGN STRATEGIES FOR THE DEVELOPMENT OF LIFE-SAVING FURNITURE SYSTEMS IN THE EVENT OF AN EARTHQUAKE

DANIELE GALLOPPO, JACOPO MASCITTI & LUCIA PIETRONI
School of Architecture and Design, University of Camerino, Italy

ABSTRACT

During an earthquake, non-structural elements, such as mobile systems, furniture and equipment that set up environments, can either become obstacles and barriers that aggravate dangerous conditions – especially for the elderly, children and those who have ambulation impairment or, on the contrary, they may represent a chance of survival in the event of collapses. This paper aims to investigate the possibilities to innovate and transform, from a structural and functional perspective, the design of furniture and mobile equipment, in intelligent safety systems that can contribute to the protection of life. In particular, the state of the art of earthquake-proof projects and products, are analyzed in a critical way both to highlight the "life-saving" implemented strategies and to bring out, at the same time, any latent needs that are not yet met but that can be developed in a future scenario. The aim of this analysis is the definition of a set of technical performance requirements, for the development of new "collaborative" life-saving concepts in the case of earthquakes. For the critical analysis of the state of the art, we carried out a series of data collection activities of patents and anti-seismic furniture products and an interdisciplinary research, concerning the health conditions and the conditions of psycho–physical stress that people undergo during a seismic event. The data collection was conducted through the analysis of the most recent scientific publications and technical sheets of existing products, and also through interviews with specific figures who were directly involved in the last seismic events that hit the centre of Italy. The paper shows the results of this research which, in turn, demonstrates how the theme of furniture design and non-structural elements for the preservation of life in the event of earthquake is current and promising; it also identifies design strategies for the development of systems of life-saving furniture, in a systemic and collaborative perspective.

Keywords: design for safety, earthquake proof furniture, furniture design.

1 INTRODUCTION

The seismic events of central Italy, especially those of 2016 that affected the municipalities of Amatrice and those surrounding the city of Ascoli Piceno, have rekindled the debate on the safety of public and private buildings in our country, especially for those that are located in historical centres. The last disastrous seismic events have highlighted, for the umpteenth time, the precariousness and inadequacy not only of the structures that constitute the heart of our cities and countries but, in particular, of sensitive buildings for communities such as schools and administrative centres. At the same time, recent telluric events have shown again the substantial inadequacy of the usual domestic practices for survival in the event of an earthquake, such as sheltering under the door lintels, whose purpose is to give greater chances of survival in case of failures and collapses, as now widely described by the theories of the "triangle of life" and by those of the "drop, cover and hold on" [1]. During an earthquake, it is not only the architectural structures that can cause serious damage to people: even the furniture systems, the objects and the equipment that set up the space can become the cause of death or obstacles and barriers which aggravate the dangerous conditions, in particular for some sensitive categories of people such as elderly, children and the disabled. On the contrary, in rare cases, the furniture can be a temporary shelter, thanks to their ability to withstand the weight of collapsing structures. We can consider these occasional shelters as

WIT Transactions on The Built Environment, Vol 189, © 2019 WIT Press
www.witpress.com, ISSN 1743-3509 (on-line)
doi:10.2495/SAFE190071

passive protection systems, as well as some devices designed for safety in other areas (airbag systems, accident prevention clothing, etc.). Their different behaviour depends on how they were conceived, designed and implemented. For many years now, there are available on the market various devices for fixing non-structural elements – such as wardrobes, electronic equipment, shelves and other potentially dangerous furniture – to the buildings' structure, so as to allow them to be secured to the walls. These are simple systems such as safety belts, metal brackets, door locking systems and anti-tipping gel pads. In 2005, a first non-structural elements study was conducted in order to provide guidelines for risk reduction and for securing the above mentioned elements through the use of these devices [2]. However these systems are effective only if placed inside buildings made with anti-seismic criteria, as the Japanese experience shows; instead they are totally inadequate, when used in a context of historic architecture such as the Italian one, which is characterised by a widespread uncertainty of the behaviour of the structures in the event of an earthquake. To date, intervention solutions have also been developed that offer greater life-saving performance, but that are also characterised by more invasive installation methods. These are hermetic structures such as bunkers and lifesaving capsules that are installed in homes and sometimes offices and public places. An interesting case is the K 107 capsule, a life-saving cell in case of natural disasters which has been developed in Mexico in 2018, in collaboration with the aerospace industry which is now in the commercialization phase.

Certainly the issue of security in the event of an earthquake, does not only concern the safeguarding of people, but also the one of cultural heritage, especially museum's works of art. In 2001, for example, the ENEA (Italian National Agency for New Technologies, Energy and Sustainable Economic Development) conducted research for the development of criteria and methods for operating interventions that are respectful of cultural values and at the same time rational and efficient, able to guarantee the reduction of foreseeable damages in the event of an earthquake. One of the strategies identified by the study and then adopted was that of integrating insulating bases to the showcases in charge of safeguarding the works of art. One of the main results of this research is a software, followed by a manual [3] addressed to the employees, with the guidelines to be followed for the protection of works of art in the event of an earthquake. A similar study but aimed at the development of life-saving furniture dates back to 2015, when four researchers from the "Forestry College of the Sichuan Agricultural University" [4] proposed a first reflection on the generation of life-saving furniture in the event of natural disasters, also spelling out some possible general design principles. To date, the state of the art and the reference literature, still appear little investigated and developed, especially under a point of view of seismic adaptation and safety of those contexts considered to be the most sensitive, such as the historical centre of areas with high seismic hazard.

Being aware that today, the processes of adaptation and securing of historic buildings in Italy, will be slow and expensive, it is necessary to envisage alternative roads to give greater safety to the communities that live in areas with high seismic risk.

One of these is the use of furniture capable of not representing an aggravation of danger but, on the contrary, act as a shelter in the event of an earthquake. Schools and public buildings represent the most critical contexts in Italy, in which the issue of passive safety provided by furniture elements can find a greater development.

2 THE REFERENCE CONTEXT

Historically, seismic engineering has focused primarily on the structural response of buildings subject to the earthquake, and on how to mitigate the damage caused to buildings. However in the last ten years, the design vision of the concept of safety within an architectural structure has greatly expanded since it has been recognized – even for non-structural elements – a strategic anti-seismic role, given that they can collaborate for the safety of the building itself and for the protection of people.

In many seismic events, it has been observed that, inadequately designed non-structural elements, can become a danger to human safety: an emblematic example is the 1995 Kobe earthquake, in which 33 people died crushed by the overturning of some types of home furnishings. However the furniture can also represent a precious opportunity for salvation, as observed in the 2017 Ischia earthquake, in which a newborn child has found vital shelter in the overturning of the cradle, which has protected him from being crushed by the rubble.

In the last years, the interest on the issue of seismic adaptation of non-structural elements has progressively grown, particularly after the Kobe earthquake. In 2008, Kimiro Meguro of the University of Tokyo conducted a first study on the behaviour of the environments for the anti-overturning system [5].

The ever increasing attention to the theme is also demonstrated by the analysis of the increase in patent applications, and by the patents granted since 1995 at an international level, relating to new furniture solutions, for safety and protection in the event of an earthquake in the domestic environment. Patent development of life-saving furniture in the event of an earthquake from 1995 until today. The following diagram (Fig. 1) shows the main CPC classifications (Cooperative Patent Classification): A47C – furniture for sitting or lying down

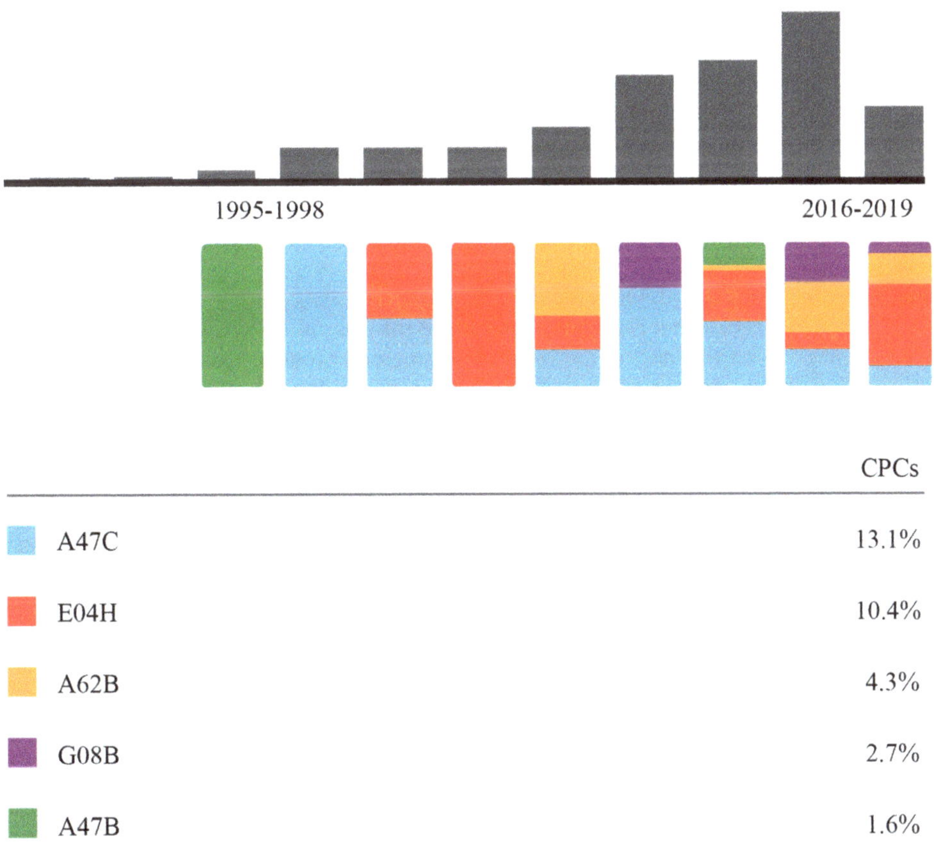

Figure 1: The main CPC (cooperative patent classifications).

and furniture that can be transformed into other types of furniture; E04H – structures for buildings to carry out special purposes (in particular E04H9/002 – earthquake-resistant structures and protection cells placed inside buildings); A62B – life-saving devices (in particular A62B31/00 – mobile shelters including respiratory protection devices); G08B – signalling and alarm systems; A47B – tables, desks, cabinets and office furniture.

To further confirm the interest on the subject of furnishing design capable of providing protection from the earthquake, in addition to the aforementioned patent phenomenon, there is a growing proposal and dissemination of competitions of ideas and projects, aimed at engineers, architects and designers worldwide. The 2016 international competition "Creation de mobilier parasismiquedesign of furniture earthquake-resistant" [6] announced by the Order of French Architects – following the Amatrice earthquake of the same year – had as objective the development of design responses for the domestic environment, able to provide shelter in the event of an earthquake. In 2017, the magazine "Eleven Magazine" for the seventh edition of its annual competition, launched the theme "Shelter 48 Emergency Life-Support Design", inviting its audience to face the subject of protecting the victims during the very first 48 hours following a natural disaster. Recently, the Italian National Association of Earthquake Engineering (ANIDIS) for its XVIII National Conference – which will be held next September in Ascoli Piceno – has included, among its discussion sessions, a session specifically dedicated to non-structural elements entitled "Industrial research on seismic performance of non-structural elements" [7]. In conclusion, until very recently, the market of products and devices for safety and protection from earthquakes – mainly of Japanese manufacture – concerned only systems of adaptation and easy installation, in order to secure traditional furniture on the walls, and reduce the possibility of overturning (i.e. metal brackets, straps, anti-vibration supports, etc.) [8]. Driven by the growing social demand for safety in the event of an earthquake, some companies in the furnishing sector, have begun to invest privately in research and development activities to implement anti-seismic functions within their products. This is the case of the movable walls of the Mangini Company [9] and of the life-saving furniture proposals of the American company LifeGuard Structures [10].

3 OBJECTIVE OF THE RESEARCH

The purpose of this research is to critically analyze the offer of lifesaving furniture solutions in the event of an earthquake, in the form of commercial products, patents, prototypes and concepts for both public and private contexts, in order to bring out the aspects of the design deemed to be of greatest interest and, in addition, to highlight any critical issues. Therefore the final objective of this paper is to bring out a series of new technical–performance requirements in a life-saving key for furniture products, aimed at the school and office sectors in the event of an earthquake. The contribution of this research to the state of the art of the technical scientific literature of reference, is represented by the methodology of analysis and critical comparison of the already developed furniture that is adopted and by the design indications that derive from it, useful to designers and furniture companies interested in the development of earthquake-proof furniture for the school and office sectors.

4 METHODOLOGY AND PHASES OF RESEARCH

The research conducted wanted to critically analyze the state of the art of life-saving furnishing projects in the event of an earthquake, for already developed public and private contexts, whether they are commercialized products, patents, prototypes or concepts, in order to bring out the design aspects considered most relevant to implement the life-saving performance in the event of an earthquake and to highlight, moreover, any latent needs not yet emerged or resolved, in relation to the same topic. The analysis of the individual furniture was subsequently summarized through the creation and use of a reasoned form, necessary to explain both the life-saving design strategies identified and the anti-seismic performances obtained. The methodology used to conduct this survey is divided into four macro phases:

1. Preliminary research on the technical–scientific literature currently available;
2. Development of a summary analysis sheet;
3. Indexing of the processed forms, in relation to the identified lifesaving strategies;
4. Summary of the results and definition of a set of technical performance requirements for the development of life-saving furniture in the event of an earthquake.

4.1 Preliminary research

In the first phase of preliminary research, a series of on-desk and infield activities have been started, in order to find the largest number of data and information on the technical–scientific literature that has already dealt with the topic that is the object of this paper, i.e. the analysis of furniture and life-saving devices in the event of an earthquake. For the search of previous patents, some specialized search engines have been consulted such as: Google Patent, Espacenet and UIBM, using the following keywords: “lifesaving furniture”, “seismic furniture”, “shelter seismic desk/table”, “Seismic furniture seismic”. The same keywords, translated into Italian, were used on the Italian patent site. The commercial products were investigated by collecting the technical data sheets provided by the companies and the multimedia material available online.

The identification, collection and indexing of case studies, lasted three months and were carried out with an interdisciplinary approach, involving researchers from the University of Camerino – coming from the two subject areas of “Industrial Design” and “Structural Engineering” – and an administrative employee, expert in “protection of intellectual property and patents”.

4.2 Development of a summary sheet

The product sheet has been developed (Fig. 2) in such a way as to simplify the operations of archiving and reading important data, i.e. those deemed to be the most consistent with the objectives of the research. The cataloguing phase saw the development of a summary sheet that provides a general overview of the product, followed by a detailed sheet that contains further information. Each sheet has a colour code that identifies the design deepening state of the product under examination: light blue for the concepts, red for prototypes, green for patents and orange for marketed products. The layout of the summary sheet is composed of four parts: the type of furniture analyzed and the commercial name are specified in the coloured bar at the top; further down on the left are the details of the company or designer (or inventor) together with the country of origin, the year of production or design in the case of concepts or the year of deposit in the case of a patent; on the right, in addition to a brief description, there are logos that identify the dimensional characteristics and the technical and life-saving performances of the analyzed product; in the central part are the images of the product and finally, at the bottom on the right there is a link to an explanatory video of the product (when provided by the companies). The detail sheet shows additional information regarding technical details and insights on technological aspects. The drawings and features of the product are described in the left column, while the reference images are shown on the right. Below the layout of product sheet with the main information and technical details is showed.

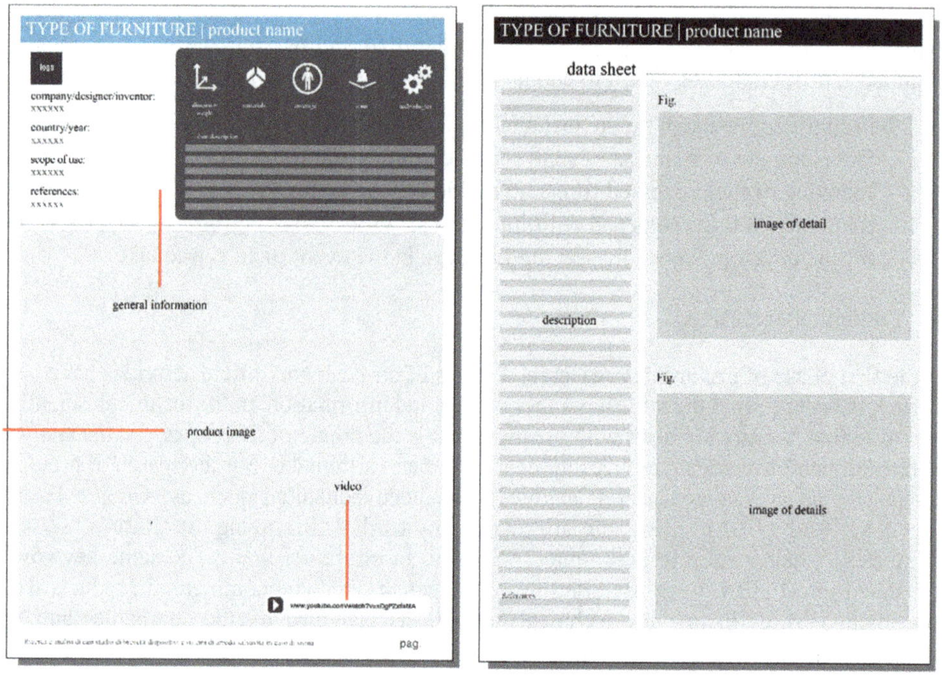

Figure 2: Data sheet developed.

4.3 Indexing of the sheets in relation to the identified life-saving strategies

In the third phase, the 42 sheets developed were divided into four macro groups: *furniture with high mechanical resistance, survival cells and capsules, transformable furniture* and *collaborating systems*. By *furniture with high mechanical resistance* (Fig. 3) we mean some types of consolidated furniture such as tables, desks, counters, etc. characterized by improved mechanical resistance to static and dynamic loads; the safety procedure consists of: seeking shelter under the furniture and waiting for help, according to the practice of survival of the "drop, cover, hold-on". Within this category, 14 products were listed.

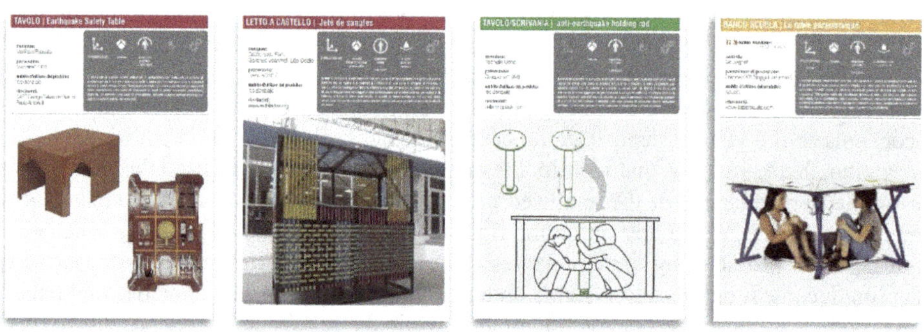

Figure 3: Four examples of *furniture with high mechanical resistance.*

By *survival cells and capsules* (Fig. 4), reference is made to all those anti seismic design solutions in which people, in order to be safe, need to reach a sort of small shelter or bunker completely closed, which, in addition to sheltering against structural collapses, it is also able to isolate the occupant from external harmful agents due to intense seismic activity: dust, gas leaks, etc. In this category, 11 products were listed.

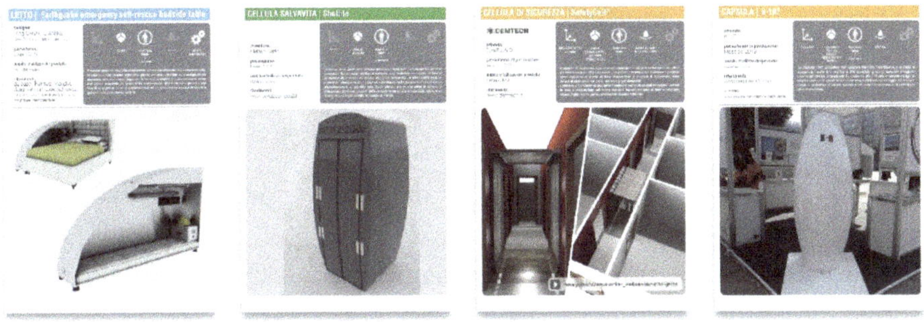

Figure 4: Four examples of *survival cells and capsules.*

By *transformable furniture* (Fig. 5) we mean those types of furniture equipped with a series of accessories that can either be activated manually or equipped with electronic sensors, which are able to trigger mechatronic systems, in turn capable of transforming the object into a bunker or into a system, that is resilient to the earthquake stresses and loads. 13 products were listed in this category.

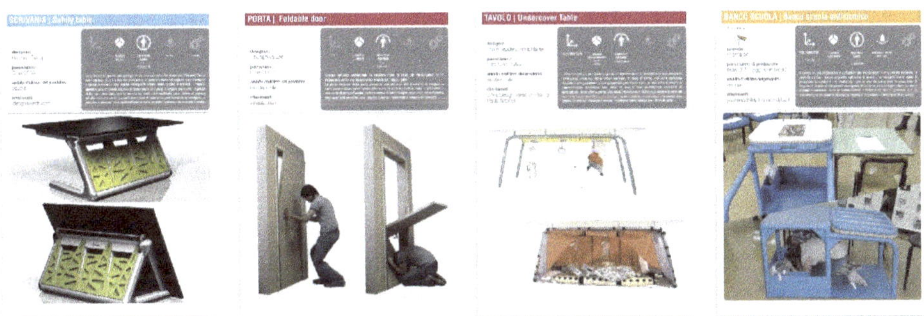

Figure 5: Four examples of *transformable furniture.*

By *collaborative systems* (Fig. 6) we mean those design solutions that work together with the building structure to secure people; in particular, by the term "collaborating", we mean the ability of a product to contribute positively to the maintenance of the static configuration of a building body, for example through a tenacious system of stiffening to be installed in a room, or a system able to follow the movements and oscillations of a building during an earthquake, thus drastically reducing the possibilities of generating barriers and obstacles. Within this category, 4 products were filed.

Figure 6: Four examples of *collaborating systems.*

4.4 Development summary sheet of the life-saving strategies adopted

Through the analysis and the study of the product data sheets, it was possible to frame at least three levels of project intervention within the various seismic hazard scenarios (historical buildings and modern buildings), which we will define for simplicity: *"light"*, *"intermediate"* and *"heavy"*. The first two concern the single piece of furniture while the third concerns the safety of a certain space of the building. More precisely, with the term *"light"*, reference is made to the use of anti-tipping and safety devices for some furniture and electronic equipment, for example: metal brackets, belts, anti-tipping gel pads, systems locking of the doors, etc. The term *"intermediate"* is used to indicate the installation of furniture and devices characterized by highly performing materials in terms of mechanical resistance and dissipative capacity, and by structures capable of transforming into "resilient" and earthquake-proof spaces (high tables mechanical resistance, "bunker" beds and "life-saving" cells). Finally, the term *"heavy"* refers to the installation of systems able to collaborate and improve the anti-seismic capacity of the building, through the realization of partially modular reinforcements and structures – generally in steel – to be placed alongside the surfaces of the room that you want to secure to generate bunker type protection spaces. This first reflection on the possible grouping of the products based on the degrees of intervention, was useful for subsequently framing within a conceptual summary scheme (Fig. 7), three project macro-objectives, concerning the great theme of the protection of people in the event of an earthquake:

1. Development of a hermetic protective space;
2. Development of an open protective system against vertical dynamic and static loads;
3. Development of a system that reduces the probability of collapse of the building structure.

The first two objectives focus on the single piece of furniture, while the third focuses on the development of a system capable of securing a specific space in the building. Each macro-objective includes a series of strategies and intervention actions that characterize the life-saving performance of a specific product; in particular the following five main strategies have been identified:

1. Increase the mechanical resistance to withstand exceptional loads;
2. Generate variable structural configurations able to withstand exceptional loads;
3. Mitigate and dissipate the stresses generated by an earthquake and/or by any exceptional loads;

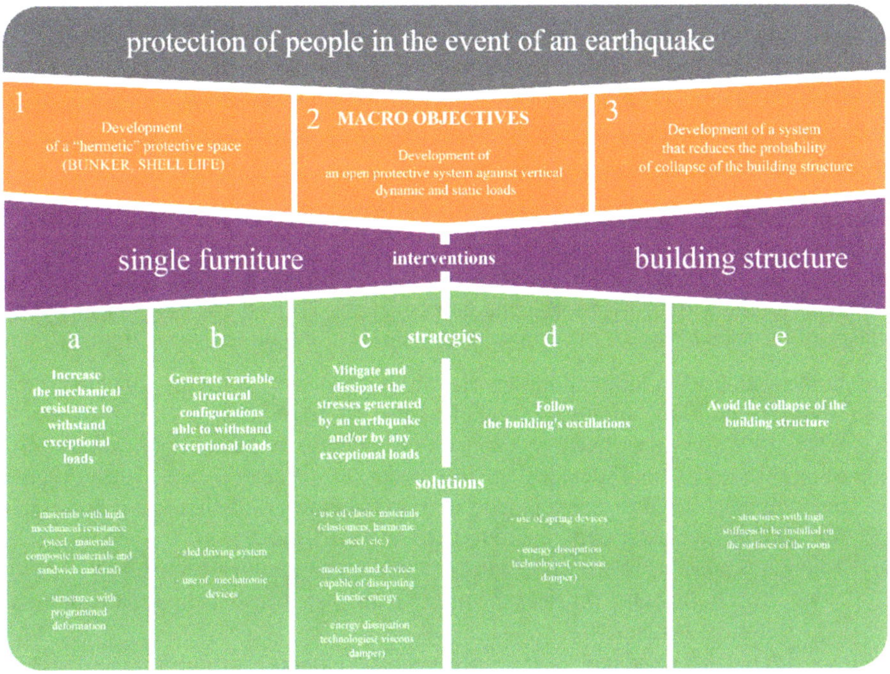

Figure 7: Summary sheet of life-saving strategies adopted.

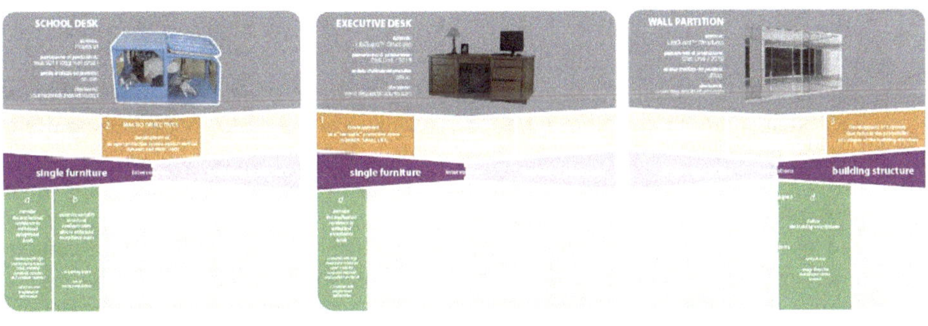

Figure 8: Examples of life-saving strategies for three case studies.

4. Follow the building's oscillations;
5. Avoid the collapse of the building structure.

On the basis of the mapped products, each strategy was able to combine some more detailed design solutions: materials with high mechanical resistance, materials and devices capable of dissipating kinetic energy, structures with high stiffness, structures with programmed deformation, etc. The tree-like scheme thus generated has therefore allowed the construction of a summary sheet of the different life-saving strategies adopted by the 42 mapped products.

This can represent a valid tool for critical inverse reading of the products, allowing, for each case study, to highlight the main objective, the context of the design intervention, the design strategy that characterizes the anti-seismic performance and the detailed solutions developed and implemented within the products. Below is showed the summary sheet of life-saving strategies adopted and its declination on three examples of products.

5 CONCLUSIONS

The research activity that has been carried out has highlighted the importance and topicality of the subject of safety and passive protection of people in the event of earthquakes and other natural disasters, leading to the identification of 42 life-saving case studies, including commercial products, prototypes, patents and concepts. From the activity of analysis and synthesis of the state of the art of the products, a series of important considerations emerge:

1. In the sector of building and industrial design, the interest for the safety of furniture within the building – non-structural elements – is booming;
2. Since 1995 there has been an exponential increase in patent applications for life-saving products and devices in the event of an earthquake, with interesting repercussions in the development and marketing of earthquake-proof types of furniture, in particular with regard to beds and work tables;
3. Today it is possible to specify at least three intervention levels for safety:
   - The "light" one: foresees the use of anti-tip devices on common wall mounted storage systems;
   - The "intermediate" one: involves the development of furniture and protection devices, resilient to dynamic and static loads;
   - The "heavy" one: involves intervention on the building envelope for the development of bunkers and life-saving cells;
4. The physical and dimensional characteristics of the cataloged products are substantially incompatible with the spaces and the precariousness of the structures of the urban fabric of the middle Italian historical center;
5. Considering the commercial landscape of life-saving products for schools, there are no design solutions for other types of products such as desks, bookcases and other storage systems;
6. Despite the considerable number of students with disabilities, "inclusive" life-saving types of furniture have not yet been developed;
7. The topic of finding people and monitoring their health status is still little investigated, and is awaiting technological development and platforms capable of coordinating and guiding the rescue forces.
8. To date there are no systems of products – neither as a concept, as a patent nor as a commercial one – that, in a widespread and collaborative manner within an environment, are able to improve the chances of survival in the event of an earthquake.

The general picture of the state of the art of life-saving products and devices in the event of an earthquake, in particular those specialized for the school and office sector, still shows some project areas that have been little investigated and are still awaiting further developments.

In the future, the different types of furniture specialized in people's protection, in addition to fulfilling their traditional peacetime functions, will have to implement a series of life-saving strategies in wartime which include:

1. New methods of collaboration between the vertical elements (e.g. wardrobes and equipped walls) and the horizontal ones (counters, desks and desks) of the "system";
2. Integration of smart systems to coordinate the rescue forces and monitor the building state of danger;
3. Inclusive solutions for sensitive subjects such as the elderly, children and the disabled;
4. Use of materials and development of design solutions for the construction of structures that are resilient and light at the same time, to facilitate the installation and application of furniture in buildings in the historic centre;
5. Use of systems for the dissipation of seismic energy, to improve the response to impacts or contribute positively to the anti-seismic capacity of the building.

The design in a systemic perspective of life-saving furniture is probably the most interesting challenge for the designers of tomorrow; it will allow us to overcome the performance offered by a single piece of furniture, while also allowing us to avoid the traditional strategy of structural oversizing, to achieve the ambitious goal of giving greater safety to communities that still live in high-seismic risk contexts and cannot rely on anti-seismic building structures for their own protection.

REFERENCES

[1] Copp, D., American Rescue Team Survival Magazine article. *The American Rescue Team International*, pp. 32–37, 2000.

[2] Wiss, Janney, Elstner Associates, *Earthquake Hazard Mitigation for Nonstructural Elements,* pp. 1–48, 2005.

[3] ENEA, *MUSEUM, manuale d'impiego,* pp. 4–71, Rome, 2001.

[4] Chen, M., Jiang, L., Liu, D.-Z. & Lyu, J.-H., *Primary Research on Emergency Self-Rescue Furniture Design for Natural Disasters,* 2015.

[5] Masatsuki, T., Midorikawa, S., Ohori, M., Miura, H. & Kitamura, H., Seismic behavior of office furniture in high-rise buildings due to long-period ground motions. *The 14th World Conference on Earthquake Engineering*, 12–17 October, 2008.

[6] Portail Humanitaire, *Le Concours International sur le Mobilier Parasismique se Poursut.* www.portail-humanitaire.org. Accessed on: 1 Feb. 2018.

[7] ANIDIS 2019, XVIII Convegno. http://convegno.anidis.it/index.php/anidis/2019 (Ascoli Piceno). Accessed on: Sep. 2019.

[8] Meguro, K., Ito, D. & Sato, Y., Efficiency of furniture overturning protection devices during earthquakes: An experimental and numerical study. *The 14th World Conference on Earthquake Engineering*, Beijing, China, October, 2008.

[9] Petrone, C., Magliulo, G. & Manfredi, G., *Shake Table Tests on Standard and Innovative Temporary Partition Walls*, Wiley Online Library, 2017.

[10] LifeGuard™ Structures, *Life-Saving Structures Multi-Hazard Protection in a Familiar Form*, pp. 2–8. www.lifeguardstructures.com. Accessed on: 3 Dec. 2018.

AUTOMOTIVE RADAR APPLICATION FOR STRUCTURAL HEALTH MONITORING

ENNIO GAMBI, GIANLUCA CIATTAGLIA & ADELMO DE SANTIS
Department of Information Engineering, Polytechnic University of Marche, Italy

ABSTRACT

Monitoring of communication and residential infrastructures, such as buildings, bridges and tunnels, has always been important to ensure the safety of citizens. In particular, it is of interest to verify the level of mechanical stress to which bridges and buildings are subjected during earthquakes, measuring the oscillations that the structure undergoes during the seismic event, in order to verify the overcoming of safety thresholds, in which stability may be impaired. In this paper the mmWave radars designed for the automotive world is taken into account, since it allows high precision measurements at a relatively low cost. mmWave radars are able to detect vibrations of the order of tens of microns and, therefore, are very useful for monitoring buildings. Furthermore, using MIMO (Multiple Input Multiple Output) mmWave radars, it is possible to implement the Beamforming technology for the spatial shaping of the radio beam, in order to detect more targets simultaneously. In this way it is possible to monitor multiple buildings or structures simultaneously, or different parts of the same structure with a single sensor. The objective of the present work is to show the applicability of automotive radar to the measurement of oscillations suffered by buildings and bridges, in order to be used as a monitoring tool in the event of earthquakes. Results of software simulations, laboratory test and real measurements on infrastructures are provided.

Keywords: mmWave radar, contactless measurements, radar sensors.

1 INTRODUCTION

Recent events in Central Italy, such as the seismic events of 2016–2017 or the collapse of the Morandi Bridge of Genoa in 2018 have increased awareness on the importance of communication and residential infrastructures monitoring. In order to ensure the citizen safety, buildings, bridges and tunnels must be constantly monitored, implementing acquisition systems supervised by artificial intelligence algorithms in order to identify any anomalous or dangerous situations. In particular it is of interest to verify the level of mechanical stress to which bridges and buildings are subjected during earthquakes, measuring the oscillations that the structure undergoes during the seismic event, with the aim to verify the overcoming of safety thresholds, in correspondence of which stability may be impaired.

For this purpose, many technologies based on contact measurements are available, that may be implemented with accelerometers [1]. In some applications, however, it is necessary to monitor the behavior with non-contact or remote methods, such as the laser or radar methodologies that provide a measure of the oscillations suffered by the structures on the basis of the analysis of the signal reflected by the structure under examination. RADAR interferometry for non-contact measurements has been applied for some time now, usually using non-commercial and often customized designed systems. In [2], [3] a radar with Stepped-Frequency Continuous Wave (SF-CW) modulation is considered, while in [4] a Frequency-Modulated Continuous Wave (FMCW) radar is taken into account. In [5] a comparison between RADAR vibration and accelerometer measurements is reported. The comparison shows how these two sensors generate signals with the same oscillation frequencies even if with different amplitudes. However, the radar highlights motion components that the accelerometer is unable to detect.

www.witpress.com, ISSN 1743-3509 (on-line)
doi:10.2495/SAFE190081

Radars have become very common in the automotive fields, as an aid to autonomous driving. These devices exhibit a typical visibility range of 150 m, very high precision and low cost thus making them very suitable for structures monitoring. The detection of vibration of the order of tens of microns can be achieved by software analysis of the phase signals. The very short wavelength of the transmitted signal allows the implementation of the MIMO (Multiple Input Multiple Output) technology, that requires the adoption of arrays of antennas. With MIMO it is possible to implement the Beamforming technology for shaping the radio beam in its form and direction according to the measurements needs. A single radar is then able to monitor multiple targets simultaneously, or different parts of them.

The objective of the present work is to show the applicability of automotive radar to the measurement of oscillations suffered by buildings and bridges, such as to be used as a monitoring tool in the event of earthquakes.

Firstly, a software simulation of an automotive radar is presented, where the reflecting signal is simulated considering an accelerometric acquisition of a real earthquakes. Then a laboratory test is showed, where the vibrating target is simulated using a metallic plane excited by an acoustic transducer, driven by a signal that reproduces traces similar to the ones due to an earthquake. Once the algorithm has been defined and validated, it was carried out an evaluation in a real context. With the aim of analyzing situations similar to those caused by an earthquake, real cases of structures subject to stress are evaluated. As a case of use the oscillation of a viaduct, solicited by heavy traffic, is analyzed.

The paper is organized as follows. After a description of the radar technology considered in Section 2, Section 3 explains the vibrational algorithm, Section 4 introduces the results of the simulations, Section 5 shows the laboratory tests while Section 6 describes the experimental evaluations. Finally, Section 7 concludes the paper.

2 THE FMCW RADAR

In the present paper a radar designed for automotive environment is considered, that operates at the frequency band of 77–81 GHz and implements the Frequency Modulated Continuous Wave (FMCW) technology. This means that the radar is continuously transmitting a signal whose frequency varies periodically. The block scheme for such a kind of radar is depicted in Fig. 1. The upper part of the figure describes the transmission subsystem, while the lower part the reception subsystem. Transmission frequency is generated by the Frequency Generator driven by a Ramp Generator, a block that provides the driver voltage according to the timing of the Timing Generator. At the reception side the received signal is mixed with the transmitted one, providing a signal whose frequency is equal to the difference between the transmitted and the received signals frequencies, being the latter one delayed by the time needed to propagate from the radar to the target, and vice-versa.

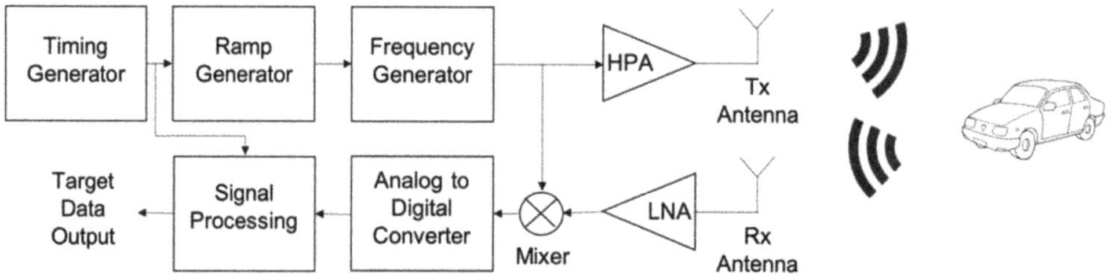

Figure 1: Principle block scheme of an automotive radar.

In FMCW radars the output frequency is linearly modulated from a minimum value f_{start} to a maximum value f_{stop} in the Chirp Time (T_{chirp}) interval, as depicted in Fig. 2; chirp sequences are grouped into frames. From the figure it is possible to note the Idle time, the interval between successive chirps, which can also be zero.

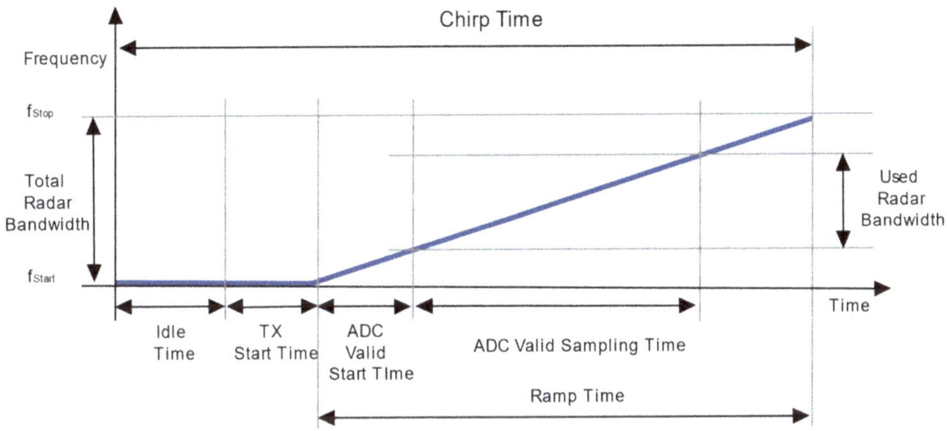

Figure 2: Timing of the Radar Tx frequency.

The output frequency remains constant at the f_{start} value during the Tx start time, needed to switch on the transmitter, while the ADC Valid Start Time defines the starting time for signal sampling, and is introduced with the aim to avoid the first part of the ramp, where the frequency evolution isn't linear. During the ADC Valid Sampling Time the mixed signal is sampled by the ADC; the total number of acquired samples is the ratio between the ADC Valid Sampling Time and the sampling time. Finally, the Ramp End Time ends the frequency transmission. As it is possible to see from the figure, the Used Radar Bandwidth is lower than the Total Radar Bandwidth, being related to the frequency evolution just during the ADC Valid Sampling Time. Radar resolution, the ability to discriminate two nearby objects, is related to the Used Radar Bandwidth and the number of acquired samples. We have to note that a at bigger values of the Used Radar Bandwidth, and then of the number of acquired samples, correspond the radar's ability to discriminate two nearby objects.

For the activity described in the present paper two commercial EVM boards from TI were used, as shown in Fig. 3. The radar board is equipped with the antennas, RF front-end, the mixer, the ADC and a processor that manages the entire board. In the board two Tx antennas and four Rx antennas are present, being each antenna a vertical array of planar antennas. The need of multiple antenna is related to the possibility offered by this board to implement the Multiple Input Multiple Output (MIMO) technology. However, in the considered case the MIMO is not necessary, since the micro-doppler information related to the vibrations that we have to measure can be obtained just from a single receiver. In spite of this, we decided to use a single transmitter ad all four available receivers.

The mixer output provides the beat frequency, which is sampled by the ADCs at a frequency that may be varied from the board configuration panel. The radar board is connected to a FPGA Lattice board through a LVDS BUS; the acquired samples are

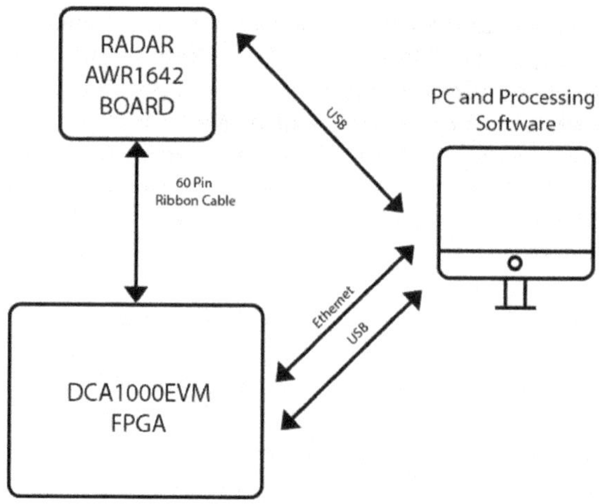

Figure 3: Radar system setup.

reorganized by the FPGA and then sent to a PC for further Matlab processing. The PC manages the configuration of both boards through USB while data are sent through Ethernet. In Table 1 the main details of Radar are reported.

Table 1: Main details of the Radar board.

Frequency	77–81	GHz
Sampling frequency	12	Msps
Maximum bandwidth	4	GHz
Maximum frequency slope	327	MHz/μs
Transmission power	12.5	dBm
Modulation type	FMCW	/
n° Tx–Rx antennas	2–4	/

3 VIBRATION ANALYSIS ALGORITHM AND CONFIGURATION OF THE RADAR

The vibration analysis was based on the work described in [6]. The vibration of the target cannot be identified from the measure of the distance due to the limited spatial resolution R_{def} of the radar, defined by the equation

$$R_{def} = \frac{R_{max}}{n^{o}_{ADC\,samples}/_{2}}. \tag{1}$$

In our case the calibration led to R_{max} of 80 m which, considering the number of samples equal to 4096, provides R_{def} of approximately 4 cm. From this result we see that it is possible to trace oscillations of a target only if the dynamic is at least of 4cm. We have to note that the oscillation of the target produces a sort of modulation on the signal phase due to the variation of the distance radar-target. The oscillation of the target gives rise to a target beat signal equal to:

$$s_{b(t)} = A \exp[j(\omega_b(t) + \emptyset(t))], \tag{2}$$

where $\omega_b(t)$ is the beat frequency due to the mean position of the target and $\emptyset$ (t) is the phase of the signal that is related to its oscillation, equal to

$$\emptyset(t) = \frac{4\pi f_C R_0 + 4\pi f_C x(t)}{c}. \tag{3}$$

The component x(t) is the generic oscillation of the target. The beat signal however does not only contain a component $\omega_b(t)$ of a single target, but also of all echoes detected. In fact, in [6] it is shown that the FFT of $s_b(t)$, fixed $\omega_b(t)$ related to the chosen target, becomes:

$$S_b(f) = \exp(j\emptyset(t)). \tag{4}$$

Referring to Fig. 4, the sampled values of the beat frequency are stored in a matrix, where in the Fast Time are stored the samples related to a chirp time, while in Slow Time are stored data related to the different chirps. In order to evaluate the phase variation of the beat signal in time, a row of the matrix is selected which is related to the distance of the target. The phase variation among the different samples on this row represents the phase change of the beat signal due to the target vibration.

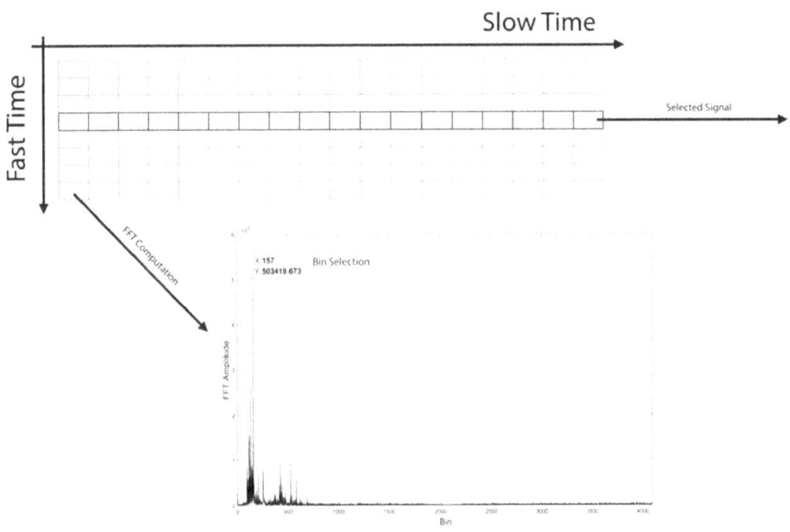

Figure 4: Storage of the beat signal samples.

The steps for extracting the phase information are:

- calculation of the FFT Range for the identification of the Target position;
- identification of the Target and its reference bin (the row on the samples matrix);
- extraction of the FFT phase of the bin;
- representation in time and frequency of the phase signal.

The radar has been configured based on two main parameters, the observation distance and the best resolution. The maximum distance R_{max} is related to the beat frequency f_{beat} through the equation

$$R_{max} = \frac{f_{beat} \cdot c}{2 \cdot p}, \tag{5}$$

where c is the speed of light and p is the slope of the FMCW signal ramp. Given that the maximum value of f_{beat} is equal to the maximum sampling frequency, only the p parameter may be chosen for calibration. The phase signal, used in the analysis, is a signal composed of a value for each chirp and therefore its sampling time coincides with T_{chirp}:

$$f_{Sampling\ FFT} = \frac{1}{T_{chirp}}. \tag{6}$$

This also determines the maximum frequency of observed vibration. The configuration we adopted tries to reduce T_{chirp} to the minimum value and to maximize spatial definition. Based on these constraints, the configuration parameters reported in Table 2 were chosen.

Table 2: Configuration parameters.

Start frequency	77	GHz
Frequency slope	11.587	MHz/μs
Idle time	100	μs
Tx start time	0	μs
ADC valid start time	0	μs
N° ADC samples	4096	/
Sample rate	12	Msps
Ramp end time	345.12	μs
Used radar BW	3998,9	GHz
Frame periodicity	60	ms

The working modulation used is FMCW with upchirp only. As result the maximum detectable distance is 80 m and the maximum detectable oscillation frequency is:

$$f_{Sampling\ FFT} = \frac{1}{T_{chirp}} = \frac{1}{Idle_{Time} + Ramp_{endTime}} = 2.246\ kHz. \tag{7}$$

4 NUMERICAL SIMULATION

In order to verify the ability of the vibration analysis algorithm to extract the evolution in time of a vibration from the radar trace, a Matlab simulation was first performed. In order to represent a realistic trend, the data of a real earthquake, which occurred in Perugia on 30 October 2016, acquired through seismic accelerometers, were taken into consideration [7]. On the basis of the acceleration data, the displacement signal, that is the entity of the vibration, was obtained through a double integration. In simulation, the displacement signal was attributed a maximum amplitude of the order of hundreds of microns, and it was assumed that the analysed target was at a distance of 3 m. The displacement signal R(t) (see Fig. 5) has a number of samples equal to 30,282 with sampling frequency equal to T_{chirp}. The beat frequency is then given by:

$$f_b(t) = 2\frac{R(t) * Ramp_{Slope}}{c}. \tag{8}$$

The value of $f_b(t)$ obtained for each $R(t)$ is used to build a sinusoidal signal $s_b(t)$ that simulates the target:

$$s_b(t) = \exp(-j\, 2\, \pi\, f_b(t)\, t). \tag{9}$$

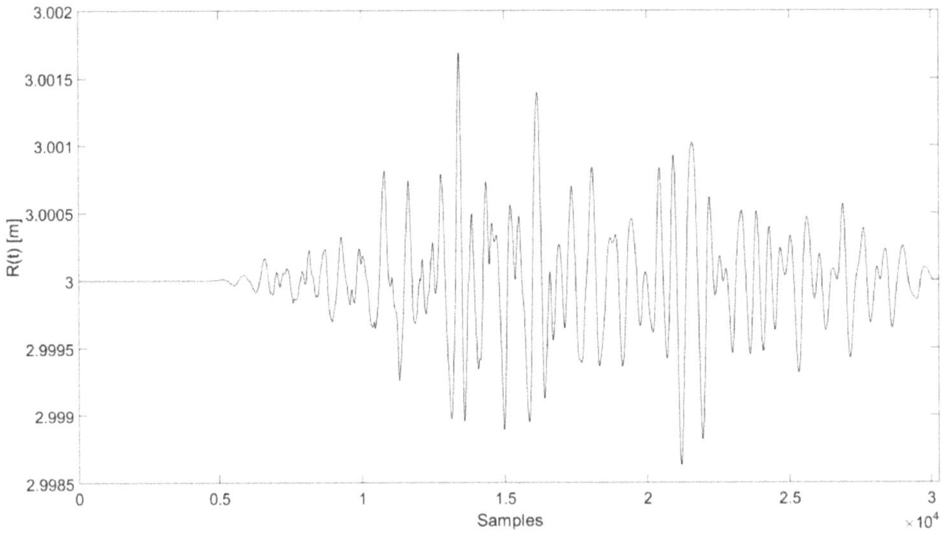

Figure 5: The vibration displacement signal.

Based on this signal, the algorithm was applied, providing the evolution shown in Fig. 6. The almost perfect coincidence of the two evolutions depicted in Fig. 5, the vibration due to the earthquake, and in Fig. 6, the phase extracted from the radar trace following the procedure described above, show the good performance of the vibrational analysis described.

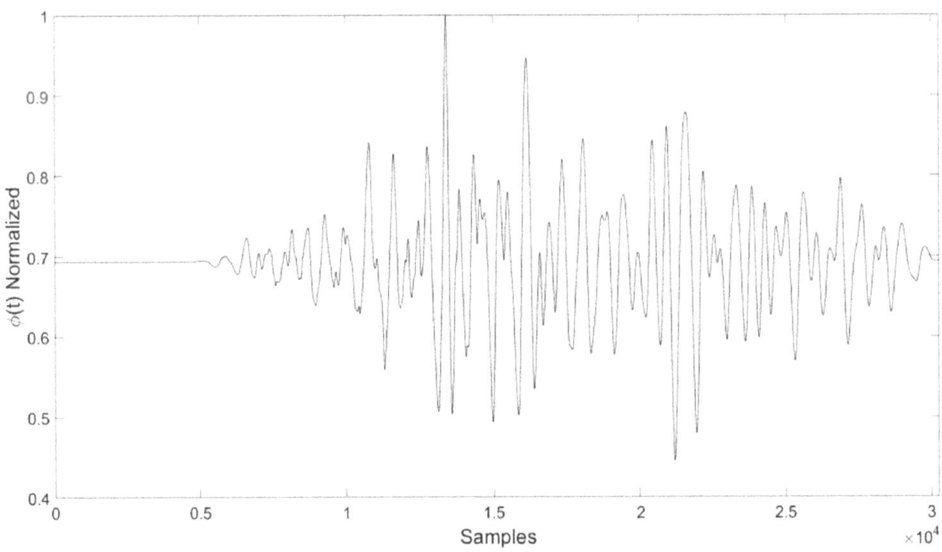

Figure 6: The phase signal provided by the vibrational algorithm.

5 LABORATORY TESTS

To validate the algorithm for vibration measurement it was decided to simulate the oscillation using an electro-acoustic transducer, fixed to a metallic board that simulates the structure under test. To excite the transducer, a train of rectangular pulses with a 50 Hz frequency and a 50% duty cycle was generated. In the measurement setup the board was placed at a known distance of about 2.59 m in order to easily locate its position in the FFT range.

The obtained results are shown in Fig. 7. In the graph of Fig. 7(a) the trend of the phase vs time is depicted, being the acquisition time of about 6 s. In Fig. 7(b) the conversion of the phase into displacement is shown, for a reduced time interval. Finally, the FFT of the displacement signal was obtained to highlight the harmonic vibration components, in order to verify the frequencies triggered by the transducer (see Fig. 7(c)). The most interesting frequency components are at 53.1 Hz and its multiples. In fact the trend of the phase signal over time makes possible to identify a periodicity of about 0.02 s. This is consistent with the type of pulse supplied to the transducer.

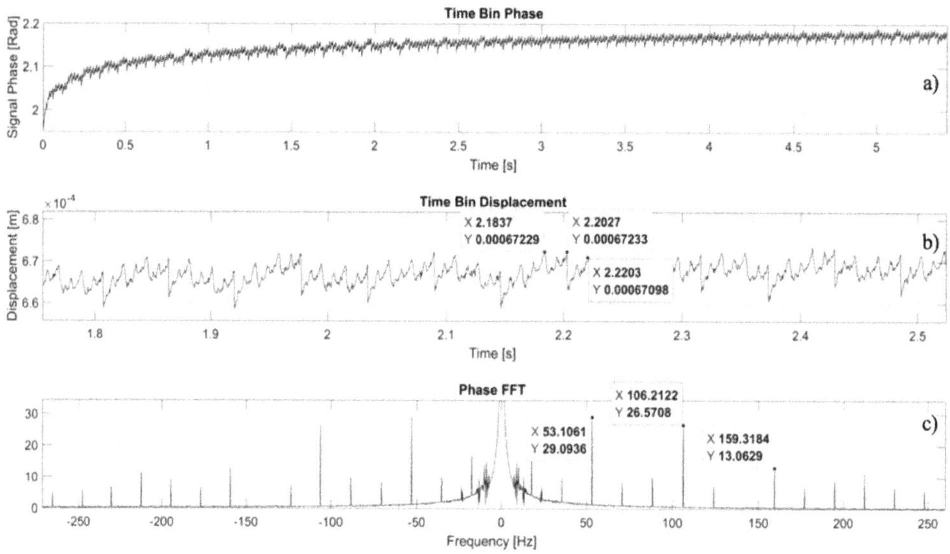

Figure 7: Laboratory test results.

6 EXPERIMENTAL EVALUATION

The last part of the work is aimed at an application in a real context, carrying out measurements on a continuously stressed structure as a motorway bridge. Sensor calibration is the same used in the laboratory environment. Two different tests were performed: in the first one the vibration of the main metal beam of the bridge was measured, while in the second one the vibration of a shelf to support the beam was considered. As introduced in Section 3 the position of the beam and of the shelf were firstly identified in the FFT Range in order to extract from the data matrices the displacement bins.

To validate the obtained results, two videos were recorded showing the passage of the various vehicles on the bridge. There is no perfect synchronization between video and radar acquisitions, but the video is still important for data validation. The results obtained are shown in the Figs 8 and 9. Fig. 8 refers to the radar pointing at the centre of the metal beam that supports the bridge. The graph shows two important oscillations spaced about 4 s apart. The analysis of the video shows two vehicles that pass on the bridge with a temporal distance of 4 s, causing bridge oscillation. Fig. 9 shows the results obtained with the radar pointing at the steel seismic restrainer of the bridge. Even in this case two important oscillations were found at a distance of about 5 s between them. The analysis of the video shows two trucks crossing the bridge with the same time interval.

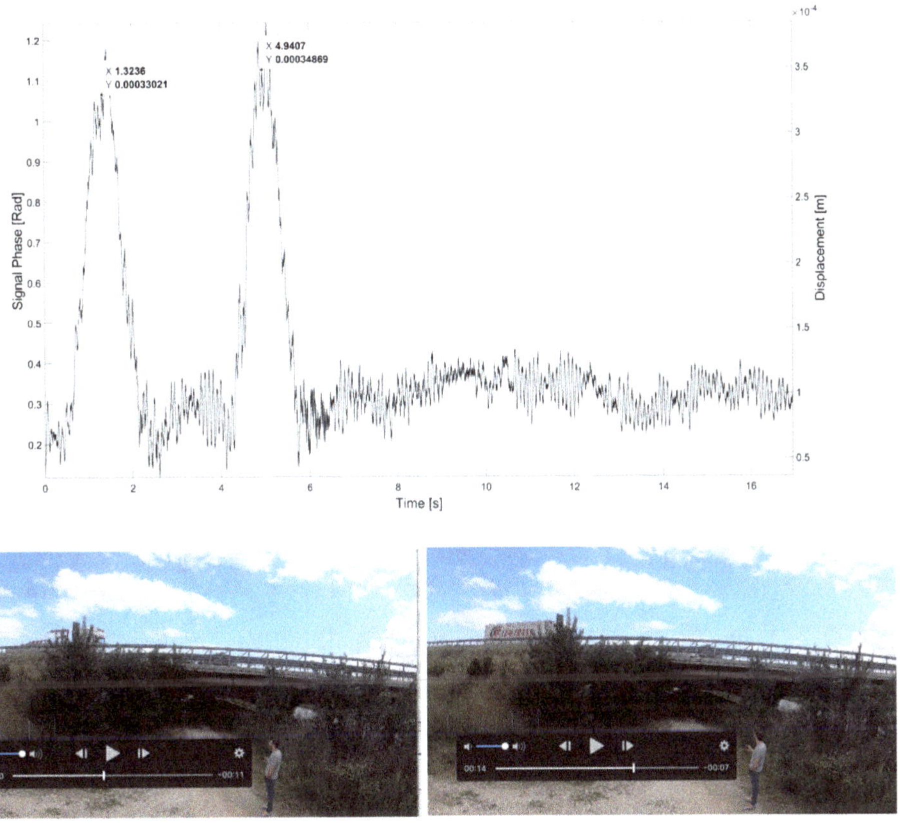

Figure 8: Vibrational test of the main metal beam of the bridge.

7 CONCLUSIONS

In the present paper the adoption of an automotive radar for the evaluation of structural vibration, as the one due to earthquake, was considered. This kind of radar is very promising for a wide range of applications, being characterized by a low cost and good performance. Thanks to its very high operational frequency, the radar and antenna dimensions are reduced, and the corresponding low value of the wavelength allows the analysis of sub-millimetre vibrations through phase analysis algorithms.

The radar performance was tested with numerical simulations, laboratory tests and experimental evaluations. In all the cases the performance was of interest, being the radar able to evidence weak vibrations too. The obtained results demonstrated that such a kind of device, developed for the automotive market, may be applied in different environment with very good performance.

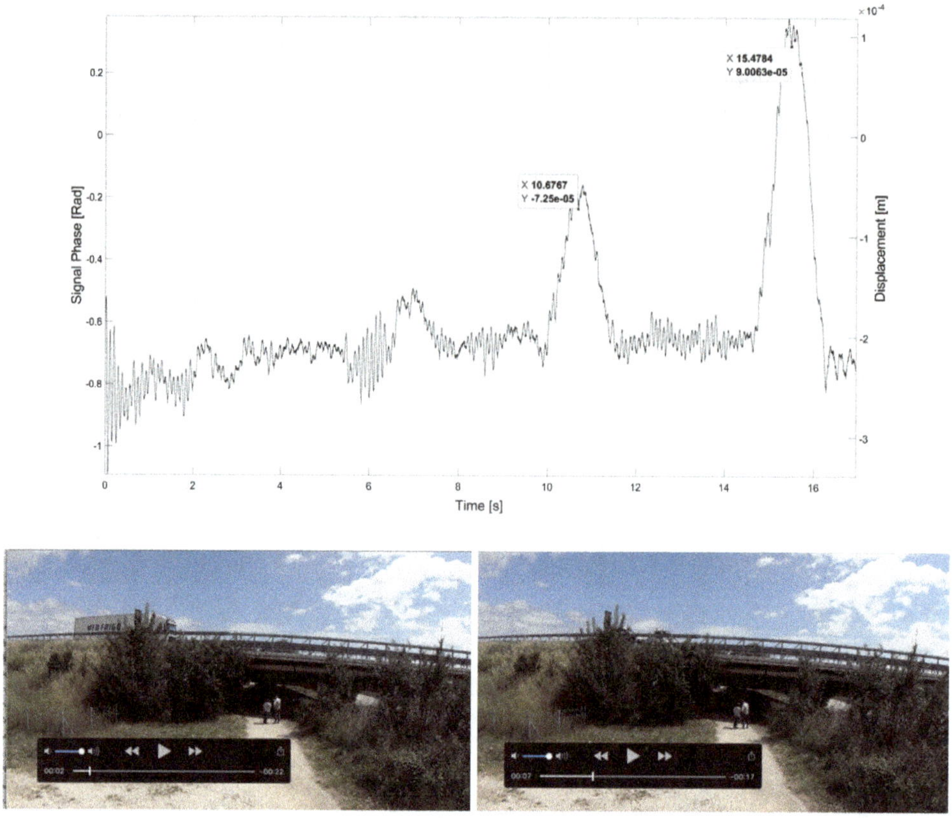

Figure 9: Vibrational test of a steel seismic restrainer of the bridge.

REFERENCES

[1] Concetti, R. et al., IoT solution based on MQTT protocol for real-time building monitoring. *Proceedings of IEEE 23rd International Symposium on Consumer Technologies (ISCT)*, 2019.

[2] Coppi, F., Gentile, C. & Ricci, P.P., A software tool for processing the displacement time series extracted from raw radar data. *AIP Conference Proceedings 1253,* pp. 190–201, 2010.

[3] Luzi, G., Crosetto, M. & Fernandez, E., Radar interferometry for monitoring the vibration characteristics of buildings and civil structures: Recent case studies in Spain. *Sensors (Switzerland)*, **17**(4), p. 669, 2017.

[4] Mei, H., Li, Y., Tian, W. & Zen, T., Weak vibration measurement based on MIMO imaging radar system. *Proceedings of the China International SAR Symposium, CISS 2018*, 2018.

[5] Negulescu, C. et al., Comparison of seismometer and radar measurements for the modal identification of civil engineering structures. *Engineering Structures*, **51**, pp. 10–22, 2013.

[6] Ding, L., Ali, M., Patole, S. & Dabak, A., Vibration parameter estimation using FMCW radar. *Proceedings of the ICASSP, IEEE International Conference on Acoustics, Speech and Signal Processing*, pp. 2224–2228, 2016.

[7] Pierleoni, P. et al., Performance evaluation of a low-cost sensing unit for seismic applications: Field testing during seismic events of 2016–2017 in central Italy. *IEEE Sensors Journal*, **18**(16), pp. 6644–6659, 2018.

PREDICTION OF SEDIMENT DISASTERS DUE TO SHORT-DURATION HEAVY RAIN IN MOUNTAINOUS AREAS BASED ON REAL-TIME RAINFALL OBSERVATIONS

TAKEYASU SUZUKI[1] & TAKUMI ITO[2]
[1]Disaster and Environmentally Sustainable Administration Research Center, University of Yamanashi, Japan
[2]Integrated Graduate School of Medicine, Engineering and Agricultural Sciences, University of Yamanashi, Japan

ABSTRACT

Owing to the global abnormal weather in recent years, short-duration heavy rain that reaches 1,000 mm in 24 h occurs locally, causing large-scale sediment disasters and flooding of rivers in mountainous areas. To help in eliminating casualties caused by natural disasters, sediment disaster alert information has been issued since 2001 based on physical modelling with an input of analyzed precipitation derived by correcting precipitation radar data with ground rainfall observations. However, in mountainous areas where sediment disasters occur, precipitation may differ greatly from the analyzed rainfall depending on the location. Therefore, we placed real-time weather sensors in sediment disaster hazard zones and predicted the occurrence of sediment disasters based on actual local rainfall observational data for the purpose of issuing evacuation information. The authors used real-time rainfall observation for measuring rainfall intensity every minute in Nishikatsura Town, which is located in Yamanashi Prefecture at the foot of Mount Fuji. Through observation, it was revealed that there was a periodic pattern of rainfall-intensity variation during short-duration heavy rainfall. The authors took into consideration this variation pattern to develop a new method of predicting rainfall at that local site by combining real-time rainfall-observation results on record for the site with corresponding data from other sites. Using this real-time precipitation prediction, it is also possible to predict the occurrence of real-time sediment disaster. In this paper, the method of analyzing the variation pattern of rainfall intensity using real-time precipitation observation data is shown, and the validity of the prediction method is verified by performing rainfall prediction for a certain observation point using rainfall-observation data from another point.

Keywords: real-time, rainfall observation, heavy rain, prediction, simulation, sediment disaster.

1 INTRODUCTION

The abnormal weather caused by global warming has brought about torrential rain that has never been experienced before, causing heavy rain-related disasters in various parts of Japan. The 2017 Northern Kyushu heavy-rain disaster caused 42 casualties (40 dead and two missing), and the 2018 Western Japan heavy rain resulted in 232 casualties (224 dead and eight missing). These disasters were caused by localized torrential rain from the occurrence of linear precipitation bands. Many casualties of heavy-rain disasters are victims of sediment disasters, which is most likely among heavy rain-related natural disasters fatal for the victims.

The most effective means to save lives from sediment disasters is early evacuation. For early evacuation, the municipalities should anticipate the occurrence of sediment disasters at an early stage, issue evacuation information, and establish a system for residents to evacuate efficiently based on that information. In Japan, according to the Sediment Disaster Prevention Act, municipalities issue evacuation information based on sediment disaster alert information announced by prefectures and the Japan Meteorological Agency. However, in the case of short-duration heavy rainfall, sediment disaster alert information may not be announced timely. In addition, even if this information is announced, there often occur cases where evacuation information is not announced before the occurrence of a sediment disaster.

WIT Transactions on The Built Environment, Vol 189, © 2019 WIT Press
www.witpress.com, ISSN 1743-3509 (on-line)
doi:10.2495/SAFE190091

Further, there is the reality that residents may not evacuate even if evacuation information is announced.

We conducted rainfall observation using real-time meteorological sensors for the purpose of supporting municipalities to issue early evacuation information by predicting the occurrence of sediment disasters due to short-duration heavy rainfall. In this paper, on the basis of the real-time rainfall-observation information from sensors installed in sediment disaster hazard zones, we present the current state of provision of information on sediment disaster risk. Then, we propose a method to use real-time rainfall-observation data from one site to estimate the short-duration heavy rainfall at another site and thus assess the risk of disaster occurrence.

2 CURRENT STATUS AND ISSUES RELATED TO EVACUATION BASED ON THE SEDIMENT DISASTER PREVENTION LAW

In the wake of the large-scale debris-flow disaster that occurred in Hiroshima Prefecture in 1999, the Sediment Disaster Prevention Act was enacted in 2001 to help in preventing future casualties in such hazard areas. In that disaster, there were 32 casualties (31 deaths and one missing) in Hiroshima City and Kure City, Hiroshima Prefecture. In order to save the lives of residents from sediment disasters, the law was formulated for the purpose of making clear areas where sediment disasters may occur, providing hazard information, establishing evacuation systems, and promoting non-structural measures such as restraint on new location of housing [1].

2.1 Outline of the Sediment Disaster Prevention Act

Prefectures designate the sediment disaster hazard zones as well as special hazard zones where disasters are anticipated to occur. On the basis of this designation, municipalities are obliged to make a sediment disaster hazard map and to make it public by providing it to residents. In the case of heavy rainfall, the prefectures jointly with the Japan Meteorological Agency announce earth and sand disaster-alert information, and on the basis of this information, municipalities issue evacuation information to residents in the identified areas and urge early evacuation. This law presupposes that community residents of the hazard zone establish a system of evacuation and evacuate voluntarily.

2.2 Issues to be solved by the Sediment Disaster Prevention Act

In Hiroshima Prefecture, a debris-flow disaster occurred in Hiroshima City in 2014, and there were 77 casualties. Although 13 years had passed since the enforcement of the law, the designation of sediment-related hazard zones had not been completed in the affected area. Even though such zones had been designated and published, 118 people died only in Hiroshima Prefecture because of the debris flow caused by the 2018 Western Japan heavy-rain disaster. Fig. 1 shows the damage caused by debris flow in Yasuura-cho district, Kure City, Hiroshima Prefecture. The sediment disaster hazard map of the same area (Fig. 2) shows that all houses are within the range designated as the hazard zone. However, the hazard map is not very useful because there is no evacuation site shown in the area. Fig. 3 shows the damage caused by debris flow in the Kawasumi district, Kumano Town, Hiroshima Prefecture. The sediment disaster hazard map of the same area (Fig. 4) shows that debris flowed out from the valley via line No. 2, which was not judged to be a dangerous mountain stream. There are a lot of issues to be solved: Sediment disasters occur before the municipalities issue evacuation information, evacuations are not performed at night even if evacuation information has been issued, and sediment disasters also occur outside areas designated as hazard zones.

Figure 1: Debris flow occurred at Yasuura-cho due to the July 2018 western Japan heavy rain disaster.

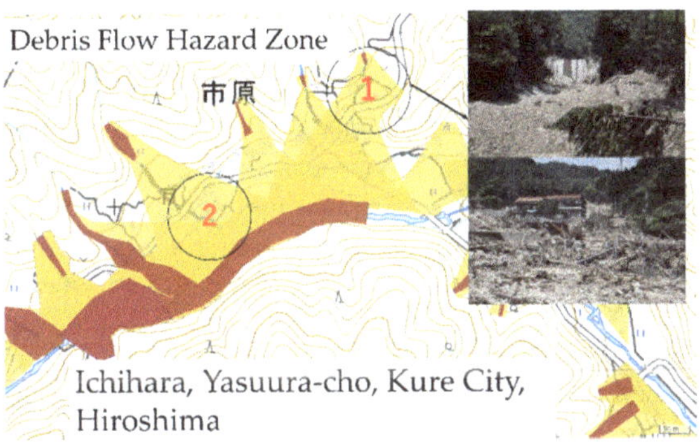

Figure 2: Sediment disaster hazard zone at Yasuura-cho, Kure City.

Figure 3: Damage due to debris flowed out from the valley line No. 2.

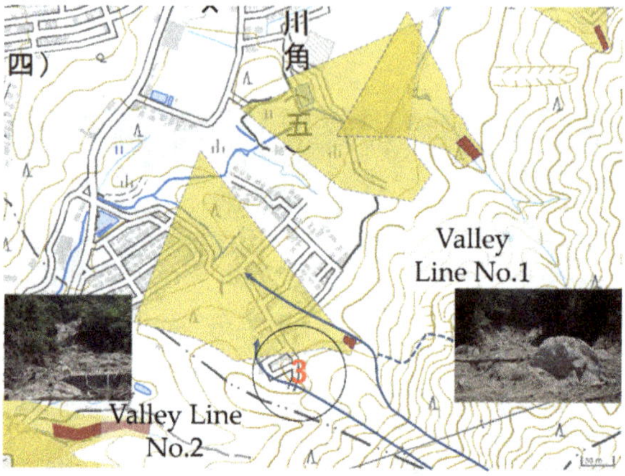

Figure 4: Sediment disaster hazard zone at Kawasumi, Kumano Town.

2.3 Issues on sediment disaster alert information

There are several issues involving evacuation based on sediment disaster alert information. First of all, the alert information is based on a critical line on a two-dimensional plot, with the soil–water index on the horizontal axis and 60 min accumulated rainfall on the vertical axis. When a point moves on this two-dimensional plane and the point is judged to have exceeded the critical line (after 2 h), sediment disaster alert information is issued. Since the soil–water index is also calculated using 60 min accumulated rainfall as input, that is the most important parameter contributing to the sediment disaster alert information.

The 60 min accumulated rainfall is the amount of precipitation analyzed by the Automated Meteorological Data Acquisition System precipitation radar and corrected by rainfall-observation results acquired on the ground. Although the amount of precipitation is announced by the Japan Meteorological Agency at a resolution of 1 km, the accuracy of the rainfall-observation points on the ground is only ~10 km. Thus, it is not suitable for local rainfall estimation. If the area where sediment disasters are likely to occur (designated hazard area) is in a mountainous area or on an alluvial fan along a mountain front, an updraft is likely to form, causing the amount of precipitation to be larger than that on the plain. It has been reported that five times the amount of rainfall in the plain has been observed in the adjacent mountainous area [2].

3 EVALUATION OF SEDIMENT DISASTER OCCURRENCE USING REAL-TIME RAIN SENSORS

The rain gauge used in this research is not a conventional tipping bucket. Rather, the rain sensors that we used measure the momentum generated by raindrops as an impact. It is known that a tipping-bucket rain gauge does not capture part of the precipitation owing to the Jevons effect, whereby air flow is disturbed near the receiver. Strong wind from the side also reduces the capture rate of rain in the receiver. However, we put more emphasis on the fact that a tipping-bucket rain gauge lacks real-time performance and is not suitable for short-duration heavy-rain measurements, as it is a measurement method that depends on temporarily storing rain. For example, when the rainfall that fell in 30 or 10 min is measured, it is a problem that a 30- or 10-min delay would occur in the evacuation judgment. In addition, to predict the occurrence of sediment disasters, we consider rainfall information at specific points in the hazard zone to be essential.

3.1 Precipitation observation using real-time rain sensor

We installed five weather sensors (Fig. 5) in Nishikatsura Town, Yamanashi Prefecture, for observation and used the data from it to provide information to municipalities in support of their evacuation judgment for sediment disasters. Nishikatsura Town requested that weather sensors be installed at five locations: two in the sediment disaster hazard zone (debris-flow hazard area), one on the roof of Nishikatsura Town Hall, and two on the bank of the Katsura River that flows east and west of the center of the town. A screenshot (Fig. 5) from the rainfall-observation website jointly implemented by the authors and Nishikatsura Town shows the actual observation information in the case of 2018 Typhoon No. 12. The number in the circle shows the accumulated rainfall for the previous 72 h, and it can be seen from the five observation points in the north and south near the mountainous area that there was already >200 mm of precipitation. By contrast, there were two to four riverside stations, and <120 mm had accumulated at two of those stations. Thus, in that typhoon, at least 66% more rainfall was observed in the sediment disaster hazard zones near mountains as on the floodplain. Sediment disaster alert information is provided for each municipality, and because the mesh resolution of the information is 5 km, Nishikatsura Town fits completely into one 5-km mesh, and uniform risk of sediment disaster occurrence is calculated in Nishikatsura Town regardless of the mountainous area and the flat area. However, as shown in the figure, it was also confirmed that the difference in rainfall amount can be double depending on location even in Nishikatsura Town [3].

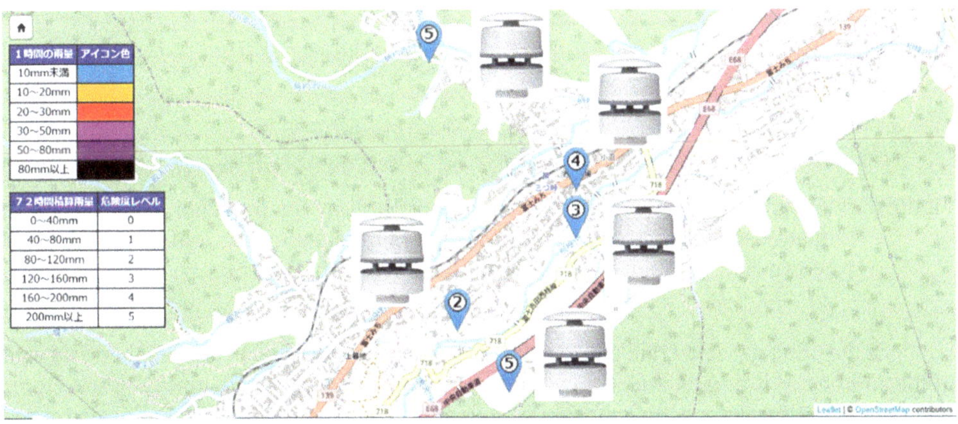

Figure 5: Sediment disaster occurrence risk information evaluated by rainfall observation using real time rain sensors (Typhoon No. 12, 29 July 2018) [3].

As the first step to evaluate the risk of sediment disaster occurrence, we created the above-mentioned website that shows real-time rainfall intensity (in mm/h) and the current status of accumulated rainfall in Nishikatsura Town at a glance (Fig. 5). This rainfall intensity is calculated by converting 1-min rainfall observed by the sensor into 60-min rainfall. The integrated rainfall (in mm) is the cumulative amount over the prior 72 h, which determines the risk of sediment disaster due to long-duration rain, and the 72-h threshold for issuing evacuation information was set at 200 mm. However, the occurrence of debris flow is predicted based on rainfall intensity, and evacuation information is issued even if the accumulated rainfall has not reached 200 mm. In addition to the sediment disaster alert information jointly announced by Yamanashi Prefecture and the Japan Meteorological Agency, Nishikatsura Town decided to issue evacuation information using the website that we created. As a result, Nishikatsura Town is able to issue evacuation information at the earliest possible time for Yamanashi Prefecture.

3.2 Verification of real time rain sensor

The second observation point from the top of Fig. 5 is the roof of the Nishikatsura Town Hall. Here Nishikatsura Town has also installed a tipping-bucket-type rain gauge to observe rainfall. When the real-time rainfall sensor was first installed there, the rainfall data from the sensor were compared to the rainfall accumulated by the tipping-bucket rain gauge to verify the observation accuracy of the real-time rainfall sensor. We found that a good correlation is obtained between the two observation results (Fig. 6(a)), if a correction is applied by taking into consideration the rainfall capture rate of the tipping-bucket rain gauge and using the wind direction and wind recorded by the real-time weather sensor.

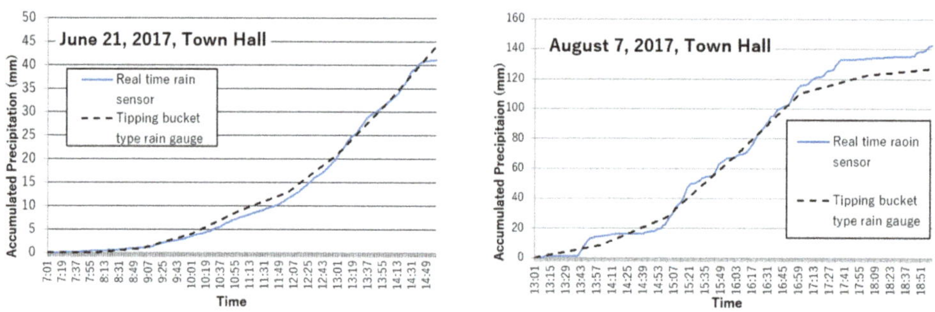

(a) (b)

Figure 6: Comparison of accumulated rainfall observed by the real-time rain sensor with that observed by a tipping bucket type rain gauge on the roof of Nishikatsura Town Hall. (a) Rain event on 21 June 2017; and (b) Rain event on 7 August 2017.

Since the tipping-bucket rain gauge is installed on the roof only 50 cm from the outer wall of the building, it is greatly affected by air flow and by structures on the roof. The structures act as obstacles, further reducing the rain-capture rate, depending on wind direction and speed. The influence becomes larger in strong wind events such as a typhoon. We compared the corrected rainfall of the tipping-bucket rain gauge with the observed rainfall divided by the capture rate and with the observed rainfall of a real-time rain sensor, taking into consideration the influence of the roof structure at the time of 2018 Typhoon No. 12 (Fig. 6(b)). Thus, we were able to verify that the real-time rain sensor is effective even in heavy rainfall accompanied by wind, although there was a rainfall event in which the cumulative observed rainfall of the tipping-bucket type rain gauge is more than 20% smaller than that of the real-time rain sensor when the influence of wind and buildings is extremely large at the time of a typhoon with extremely strong wind.

3.3 Periodic characteristics of short-duration heavy rain

During the past two years' observation period, sediment disasters did not occur in Nishikatsura Town, but debris flows occurred in nearby communities. The rainfall-intensity time history for the short-duration heavy rain observed in Nishikatsura Town, when it caused the debris flows in Ootsuki City 20 km away from Nishikatsura Town, was displayed at 1-min intervals, we found that there were peaks with a predominant period of 10–20 min. Fig. 7 shows a short-duration heavy rainfall event of 8 August 2018. Others also have noted such periodic characteristics from the measurement of rainfall intensity at 1-min intervals. Yamada et al. reported a similar rainfall pattern by observation with a special rain gauge that can monitor rainfall at 1-min intervals [2]. The 1-min-interval observations revealed that some instantaneous high-intensity rain had not been observed in short-duration heavy rain recorded at longer intervals. Because such high-intensity heavy rainfall may cause sediment runoff from the top layer of slopes and trigger debris flow, this phenomenon must also be considered a mechanism of sediment disaster occurrence.

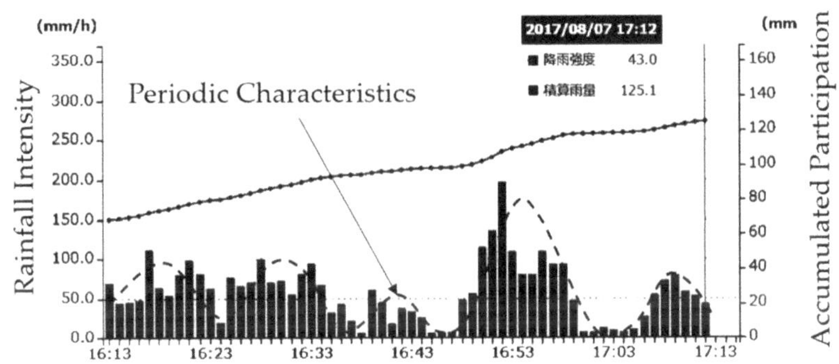

Figure 7: Periodic characteristics of rainfall intensity observed by real-time rain sensor.

By calculating a moving-average of rainfall-intensity time-series data and then subtracting it from the time-series data, temporal variation of rainfall intensity can be obtained. The low-pass filter can be adjusted by changing the width of the moving average. We determined the temporal variation using a width of 5 min, which yielded five data points. Fig. 8 is time series data of the rainfall intensity on 7 August 2017, and Fig. 9 is a result of calculating the moving average of rainfall intensity data shown in Fig. 8. Fig. 10 shows the temporal variation of rainfall intensity data shown in Fig. 8. Fourier spectrum of the temporal variation is illustrated in Fig. 11, where periodic characteristics can be clearly seen in the variation of rainfall intensity, with a predominant period of 15–20 min.

4 PREDICTION OF FUTURE RAINFALL USING AMPLITUDE AND PHASE CHARACTERISICS OF RAINFALL INTENSIY TIME HISTORY

4.1 Method of prediction of future rainfall

In earthquake engineering, there is a method of simulating earthquake motion fitted to a response-spectrum design using the phase of strong motion [4]. By this method, input earthquake motions of various phases are designed using strong-motion records of large earthquakes observed in various locations. We propose a method to forecast rainfall at a

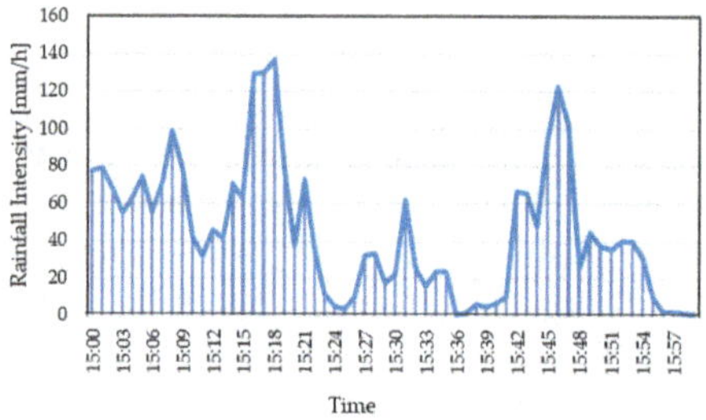

Figure 8: Rainfall intensity time history.

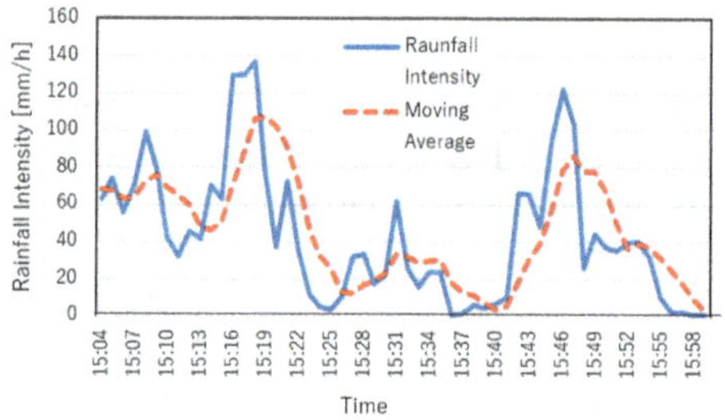

Figure 9: Moving average time-series.

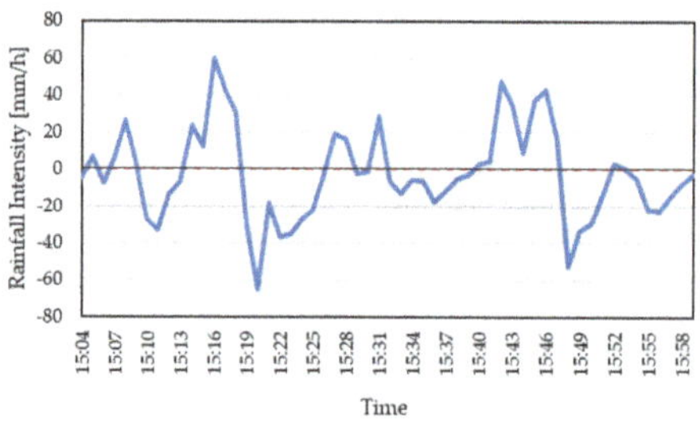

Figure 10: Temporal variation.

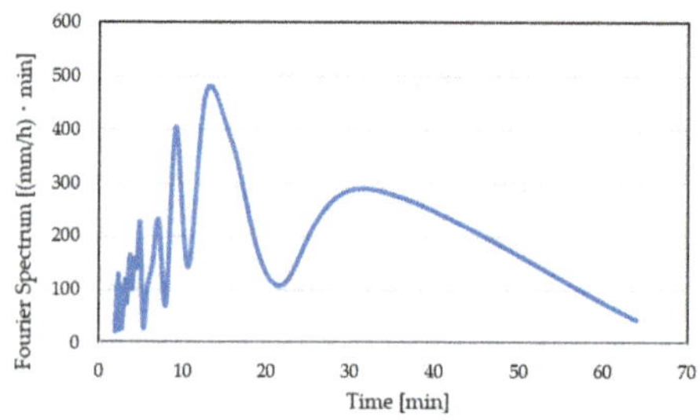

Figure 11: Fourier spectrum of the variation.

certain observation point by applying essentially the same method. In this case, when a rain cloud that causes heavy rainfall moves from observation point A to observation point B and when a high correlation can be observed between the rainfall-intensity time histories at the two observation points, it is possible to predict future rainfall at observation point B according to the following procedure, using the phase characteristics at observation point A and amplitude characteristics at observation point B.

1. Determine the variation component of rainfall intensity time history at both observation points.
2. Calculate the time difference T_d (in minutes) between observation points A and B from the cross-correlation function of both observation points.
3. Perform Fourier transform of the variation component at both observation points to obtain the Fourier amplitude and phase angle.
4. Using the amplitude characteristics (amplitude spectrum) of the variation component at observation point B and the phase characteristics (phase angle) of the variation component at observation point A, create the Fourier coefficient C_k as shown in eqn (1):

$$C_k = \frac{F_k}{T}(\cos\varphi_k + isin\varphi_k), \tag{1}$$

where, C_k denotes Fourier coefficient of the predicted temporal variation x_m, F_k the Fourier amplitude spectrum of the k-th element of the temporal variation at Point B,

T time duration of the temporal variation and φ_k the phase angle of the k-th element of the temporal deviation at Point A.

5. Determine a new simulated time-series component of the variation x_m by performing an inverse Fourier transform of the Fourier coefficients in eqn (1):

$$x_m = \sum_{k=0}^{N-1} C_k \, e^{i\left(\frac{2\pi k m}{N}\right)} \quad m=0,1,2,\ldots,N-1. \tag{2}$$

6. Determine the time series of simulated rainfall intensity by summing the time-series component of the simulated variation and the moving average at observation point B.
7. Use the time component T_d from the tail end of the simulated rainfall-intensity time series for the future-prediction time component T_d at observation point B.

4.2 Verification of the method

Using the above proposed method, we developed the rainfall-prediction program and performed rainfall forecasting using rainfall-intensity data from two observation points observed on 23 October 2017 in Nishikatsura Town. Rainfall intensity after 0:20 was forecast using the Fourier amplitude spectrum of the time-series component of the variation from 23:20 at the Citizen's Ground observation station (Point B, the southernmost station) and the phase angle of the time-series component of the variation from 0:20 at Irita observation station (Point A, the northernmost station). Fig. 12 compares the predicted rainfall intensity (red line) with the actual observed rainfall intensity (blue line). Assuming that the phase difference between the two points was 4.5 h, we carried out the above forecast of rainfall intensity. Because the actual distance between the two observation points is about 3 km and the phase difference (Td) is only 2 min, it is natural that they agree well. However, the validity of the proposed method was demonstrated. Fig. 13 compares the predicted cumulative precipitation (red line) with the actual observed cumulative precipitation (blue line). Similar comparisons are made with other rainfall events as shown in Figs 14 and 15.

When the distance between two observation stations exceeds the range of continuous movement of the rain cloud, such as when a cumulonimbus cloud occurs locally, over-estimation or under-estimation of rainfall occurs. For accurate early warning, therefore, it is

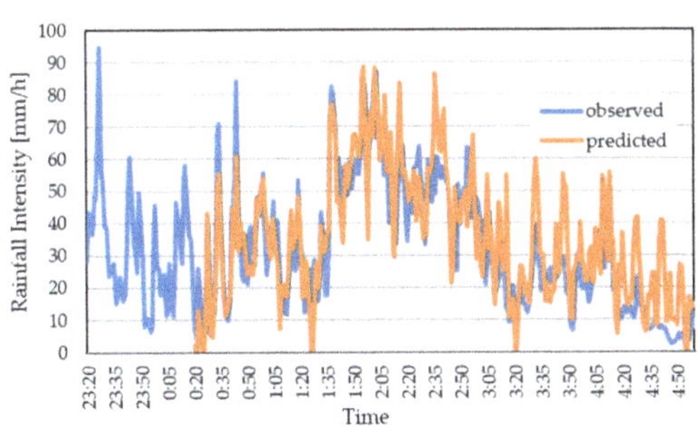

Figure 12: Comparison of observed and predicted rainfall intensity time history (23 October 2017).

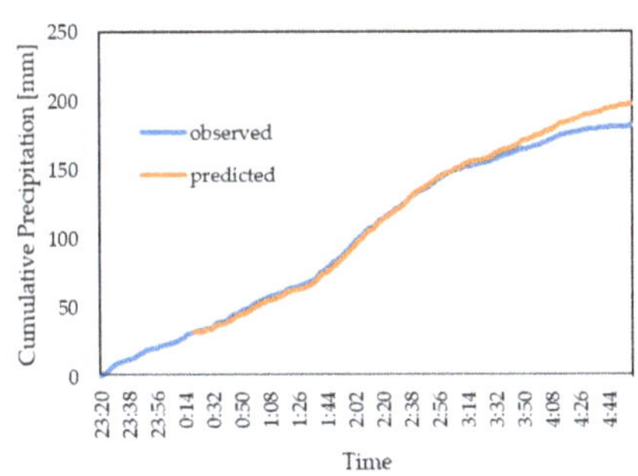

Figure 13: Comparison of observed and predicted values in cumulative precipitation (23 October 2017).

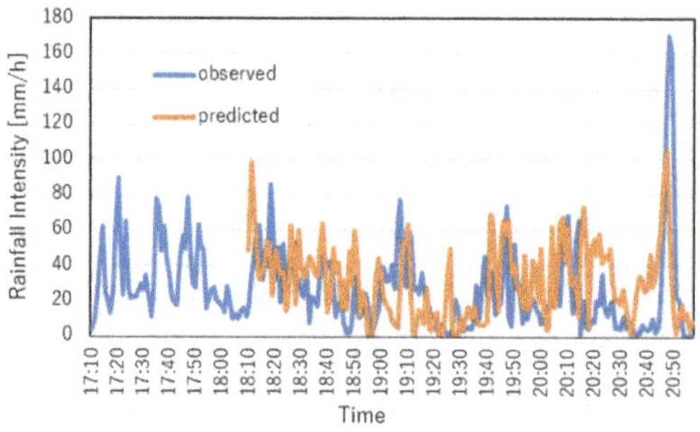

Figure 14: Comparison of observed and predicted rainfall intensity time history (28 July 2018).

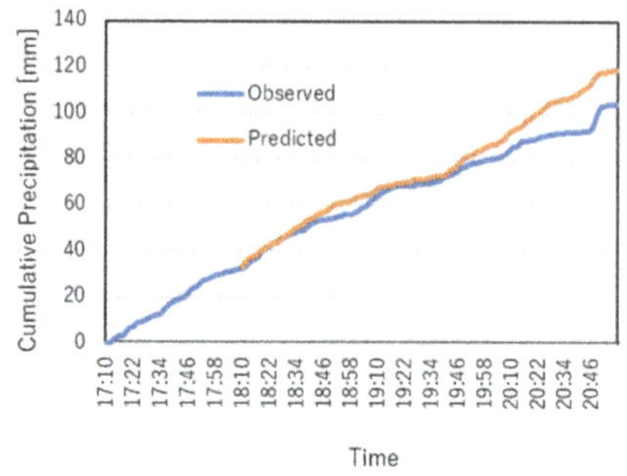

Figure 15: Comparison of observed and predicted values in cumulative precipitation (28 July 2018).

necessary to build a system that predicts rainfall by arranging rainfall sensors continuously so that movement of not only wide-area rain clouds but also local rain clouds can be taken into consideration.

5 CONCLUSIONS

By installing a real-time rainfall sensor in a sediment disaster hazard zone, it is possible to evaluate the sediment-related-disaster risk at specific points and in real time with high accuracy. Furthermore, we showed that it will be possible to forecast rainfall and sediment disaster occurrence by evaluating autocorrelation and cross-correlation of rainfall intensity between rainfall-observation stations in real time. When rainfall observation is carried out at specific points and in real time in the sediment disaster hazard zone by adjacent municipalities, it is possible to forecast rainfall of ~1 h using the cross-correlation between observation points regardless of municipal boundaries. The authors have proposed a method of forecasting rainfall, focusing on the periodic characteristics of short-duration heavy rainfall. It seems possible to forecast rainfall and sediment disasters by applying machine learning, even in cases where the periodic characteristics are not remarkable and are irregular.

REFERENCES

[1] Disaster Management Bureau, *Flood and Sediment Disaster*, Cabinet Office, Japan, 2018.

[2] Nagata, Y., Yoshimi, K., Nagashima, K. & Yamada, T., A study on the short-term characteristics of rainfall phenomenon. *Proceedings of Japan Society of Civil Engineers, G*, **71**(5), pp. I277–I282, 2015. (In Japanese.)

[3] Suzuki, T., *Disaster Management and Engineering*, Riko Tosho Publishing Company, pp. 135–136, 2019. (In Japanese.)

[4] Suzuki, T., Adachi, Y. & Tanaka, M., Application of microtremor measurement to the estimation of earthquake ground motions in Kushiro City during the Kushiro-oki earthquake of 15 January 1993. *Earthquake Engineering and Structural Dynamics*, **24**, pp. 595–613, 1995.

IMPACT OF SELF-DRIVING AND CONNECTED VEHICLES ON EMERGENCY RESPONSE: THE CASE OF THE USA AND IMPLICATIONS FOR ITALY

AUSTIN W. OBENAUF[1], REGINALD R. SOULEYRETTE[1],
ROBERT M. KLUGER[2] & ANTONIO PRATELLI[3]
[1]University of Kentucky, USA
[2]University of Louisville, USA
[3]University of Pisa, Italy

ABSTRACT

Studies have shown that mortality rates are directly correlated with emergency response times. Cardiac arrest events, trauma, and stroke are among the most time dependent. Response time, in turn, is greatly affected by traffic and procedures currently required at intersections. New technology promises more efficient flow at these locations. In particular, connected and autonomous vehicles can contribute to reductions in response times and resulting reductions in mortality. The purpose of this paper is to provide a framework and a tool for analysis of these improvements and reductions. Application of the framework for Lexington, Kentucky, USA indicate a potential three-minute decrease in response time with subsequent reduction in fatalities. Extrapolated to the USA, the savings are potentially quite significant. The paper goes on to speculate on implications for Italy.

Keywords: connected and autonomous vehicles, emergency response times, virtual emergency lanes.

1 INTRODUCTION

Connected and autonomous vehicles (CAV) have been regarded as the best response for eliminating human error and saving lives on roadways. Although the technology includes a safety benefit to decreasing roadway fatalities, what other safety benefits can the technology provide? According to the Center for Disease Control and Prevention (CDC), the top causes of death in the United States for 2016 include heart disease, cancer, accidents, chronic lower respiratory diseases, stroke, Alzheimer's disease, diabetes, influenza, nephritis, and intentional self-harm [1]. The types of deaths are correlated with response times, as some are time-sensitive and require immediate medical attention. Pre-hospital fatalities make up 16% of all fatalities in the United States [2]. Removing the immediate fatalities from impacts, most pre-hospital fatalities occur from longer emergency response times. Emergency dispatchers today are equipped with live location data of the emergency units, which allows for the closest unit to respond to a call [3]. Even though dispatching the closest unit available will save time in responding, human drivers and EMS response procedures still slow the EMS unit down, whether for an inattentive driver or an intersection ill-prepared for an emergency vehicle approach. Human interference to emergency responders poses the question, how can CAV technology eliminate human error and decrease emergency response times, increasing the survival rates of time-sensitive patients in urban areas?

As of 2018, the average emergency response time is over nine minutes [4]. Faster response times are proven to provide higher survival rates [5]. Time-dependent emergencies make up 25% of all medical calls across the nation, including 9.7% being the most time-sensitive call, cardiac arrests [6]. Cardiac arrest mortality rates increase 7% to 10% for every untreated minute [7]. Other time sensitive cases have a 1% to 2% increase in mortality rate, but the largest spikes in mortality rate occurs at the five-minute mark and the eight-minute mark [8], [9]. By reducing average emergency response times to under eight minutes, cardiac arrest

WIT Transactions on The Built Environment, Vol 189, © 2019 WIT Press
www.witpress.com, ISSN 1743-3509 (on-line)
doi:10.2495/SAFE190101

mortality rates would be reduced by around 8.5% and other time-sensitive cases would be reduced by 1.5%.

Vehicle-to-infrastructure (V2I) technology can enhance preemption at signalized intersections. Pre-routing of the emergency vehicle could enable signals to have more accurate preemption, reducing or eliminating stops at signalized intersections. In one study, connected technology was shown to provide a 34% decrease in emergency response times [10]. Vehicle-to-vehicle (V2V) technology allows for drivers to know where the emergency vehicle will be.

Unfortunately, the normal procedure, for drivers is to move over to the right side of the road, is not always possible. Autonomous vehicle programming could replace the normal procedure of human vehicles by clearing a path on roadways for emergency vehicles through a virtual emergency lane (VEL). By having clear path to the scene in advance, emergency vehicles will have minimal delay due to congestion, reducing the average response time and saving more lives for time-sensitive cases, particularly in urban areas.

To show how CAV technology can reduce emergency response times, a Microsoft Excel® spreadsheet tool called ddEMSCAT (data driven Emergency Medical Services with Connected and Autonomous Technology), was created to implement the developed framework and estimate the benefits CAVs can provide. ddEMSCAT utilizes data, user inputs, and the framework developed to forecast mortality rates under different levels of CAV technology and market penetration. The framework is demonstrated on Fayette County, a mostly rural county that includes an urban area in the city of Lexington. The tool can be used by urban area governments to inform policy and ensure the technology's potential is achieved.

In many ways, EMS in Italy and the USA are similar. Standards of care and response times compare well. Eight to nine minutes response time to life threatening emergencies is a common standard in the two countries. However, goals in some European countries greatly exceed this standard response, with 4 minutes, 45 seconds being the standard of response time in Copenhagen, Denmark. Dispatch areas (catchments), range from 30,000 persons (Cyprus) to 2.5 million persons (Portugal). Dispatch areas (size) range even more widely, from a low of 250 sq. km. in Cyprus to 50,000 sq. km. in Estonia. Italy records 103 EMS services, each serving and average of 550,000 persons with catchment area average size of 3000 sq. km. [11]

2 METHODOLOGY

CAV technology presents the potential to drastically reduce response times by allowing emergency vehicles to better navigate congested urban areas. A data-driven framework is developed in this study to estimate how different levels of CAV technology can decrease emergency response times and lower mortality rates in emergencies. Data are acquired from NEMSIS (National Emergency Medical Services Information System) and KCTCS (Kentucky Community and Technical College System) for Fayette County, Kentucky EMS calls for January 2019 [12], LFUCG (Lexington-Fayette Urban County Government) addresses, roadways with speed limits, fire station locations and trauma center locations [3], KYTC (Kentucky Transportation Cabinet) HIS (Highway Information System) roadway types and traffic [13], and Google Maps typical traffic congestion for roadways [14].

Lexington-Fayette EMS run logs are joined to KCTCS records street addresses associated with each EMS call. These are next joined to LFUCG data including street addresses, all roads (with speed limits), fire station points, and trauma center points. Run times vary by congestion level and roadway type. EMS miles traveled by road type must be estimated, so an equation is defined to estimate run times by road type. Average congestion level by percentage is used to calibrate the travel time equation [14]. These estimations are then used to compare the response time under various degrees of market penetration of CAVs. Once the change in response time is calculated, effect on mortality is computed.

2.1 Data

An open records request was sent to the Kentucky Emergency Medical Services Information System [12] to retrieve data on typical EMS runs in Fayette County, Kentucky. The request featured three types of data, including times, locations, and other. The data requested had to follow the restrictions under the Health Insurance Portability and Accountability Act (HIPAA) and did not provide personally identifiable information. The complete data received included: call time, dispatch time, unit en route, unit arrives at scene, unit departs scene, unit arrives at emergency department, origin of dispatch, patient location, location of emergency, indication of patient transport to emergency department, status of light usage to and from scene, and type of injury or call.

After the approval of the data request, all EMS run data from January 2019 in Fayette County was received. The Fayette County data contained 3,995 total records, with 1,303 non-hospitalized runs and 2,692 hospitalized runs. Although the data included valuable information for this project, 371 records were either incomplete or non-emergency, and most of the records did not have an origin of dispatch listed for the emergency vehicle. An automatic vehicle locator enables the dispatch of the nearest ambulance for a call. However, the originating location cannot be determined without making an assumption. Therefore, records without starting points were assumed to originate from the nearest fire station. The NEMSIS and LFUCG data sets were joined within ArcMap GIS, which readily enabled the computation of EMS route distances.

Congestion levels were approximated for different times throughout the day using Google Maps time of day estimates [14]. Google uses three different colors to indicate traffic levels, including green for light congestion (> 50 mph for freeways), yellow for medium congestion, and red for heavy congestion (< 25 mph) [15].

2.2 Virtual emergency lanes

Typical driving conditions are less than ideal for emergency vehicles. In the USA, most drivers follow the laws and cede right of way emergency vehicles. However, some do not comply in time to prevent delay. By eliminating human indecision and error and incorporating advanced knowledge of emergency vehicle routes, CAVs could allow emergency vehicles to respond and drive to calls more quickly and safely than today. The concept advocated in this study is that of the VEL. As moving to the right side of the road is the best solution with human driven vehicles, CAVs can analyze the situation and create space to augment or facilitate current rule following. The VEL potentially creates more space and allows EVs to pass with less impedance. With heterogeneous or fully autonomous fleets, connected vehicles will be able to create more space for EVs as compared to existing human driven non-connected systems. The two factors most impacting the performance of VELs are roadway type (geometry) and market penetration of CAVs. In this study, roadway types include freeways, five or seven lane roadways with center two-way left-turn-lane (5/7 CTWLTL), four lane roadways with left-turn lane at intersections, four lane roadways without left-turn lane at intersections, and two-lane road without left-turn lane at intersections. Three CAV market penetration scenarios are studied: 0%, 50%, and 100%.

EVs do not typically use freeways and interstates for response routing in urban areas. However, some EVs do make use of them, especially for non-emergency hospital transports or longer distances to a scene. The heaviest traffic congestion generally occurs on major arterial roadways, such as a divided roadway with barriers and left-turn lanes at intersections or divided roadways with a two-way left-turn lanes (TWLTL). The current study included two major Lexington, KY five/seven-lane roadways with dense traffic and multiple signals (South Broadway Street and Nicholasville Road).

2-Lane roadways with a TWLTL are similar to 4-lane roadways with TWLTL, but congestion is usually lighter, and space is limited for EVs. The roadway consists of a bulb to create enough space for two lanes at the intersection, while the rest of the roadway only contains two lanes throughout. Due to the usual designation of two-lane roadways as minor arterials or collectors, the signals associated generally give more green time to the cross street, creating more red time in the direction of the two-lane. CAVs have potential to improve two-lane roadways with left-turn lanes as well.

The final roadway type is a two-lane road without an LTL at intersections, usually found in suburban areas and neighborhoods. Considered as collectors or local roads, the higher ratio of green time generally goes to the cross street, limiting the amount of green time for the two-lane without left-turn lane roadways. The shoulders, if any, are generally narrower on these streets as well, requiring vehicles in both directions to leave the right-of-way or run a signal to accommodate an EV. Human drivers do not know the intended route of the EV, making communication difficult. With communication, CAVs will be able to make a more appropriate response to clear the roadway for an EV. Fig. 1 displays three scenarios for 2-Lane roads without a left-turn lane. The scenarios depict three levels of market penetration of CAVs and indicate the friction with human drivers versus the ease of coordination possible through VELs.

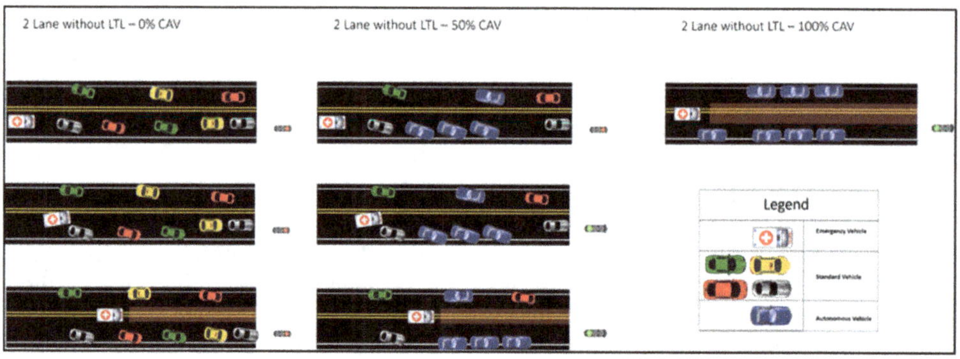

Figure 1: 2-lane roadway scenarios with virtual emergency lanes.

2.3 ddEMSCAT

To forecast the number of potential lives saved due to quicker response times enabled by CAVs, the research team developed ddEMSCAT (data driven Emergency Medical Services with Connected and Autonomous Technology). This spreadsheet-based tool combines data from EMS services, roadway characteristics, and forecasts of CAV market penetration. This section describes the framework and the implementation of the tool.

The primary contribution of ddEMSCAT is the travel time estimation of EMS through various levels of CAV market penetration. ddEMSCAT computes EMS travel time and number of patients affected, as shown in the flowchart of Fig. 2. Although the tool is data driven, certain assumptions must be made, especially as related to the market penetration of CAVs. Using Lexington EMS data from January 2019, an example application of the tool calculates time savings and potential number of lives saved.

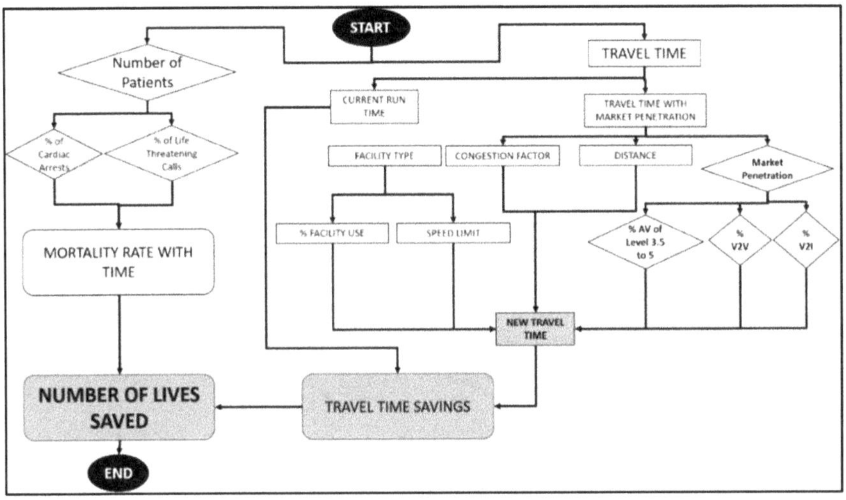

Figure 2: ddEMSCAT flow chart.

As the benefits of CAV manifest via reduced run time, calculations are made for various types of roadways, times and origins/destinations of emergency calls. Using typical run information ddEMSCAT can approximate the response times of today. For each run, a set minimum amount of time is required to initiate the run as well as to exit the vehicle and arrive on scene. From the Lexington data, this set time is taken as 1.2 minutes. This constant allows for one minute to leave the station after reporting en-route, and 0.2 minutes (12 seconds) to slow down at the scene and reporting "at-scene." Also factored into the minimum response time is travel time at free flow speed, which is taken as the distance to the scene dividing by the speed limit plus 10 mph, the maximum permitted EV speed. See eqn (1)

$$Open\ Road\ Response\ Time\ (min.) = \frac{L}{SL+10mph} x\ 60 + 1.2, \qquad (1)$$

where L = Length (miles) and SL = Speed limit (miles per hour).

Under light traffic conditions, Eqn 1 remains the same, but medium and heavy congestion will cause slower response times. Since Google Maps (source of ddEMSCAT typical travel times) provides a range, rather than an exact speed, the midpoint of the range is used to represent each traffic condition. For medium congestion, 37.5 mph is used to represent speed of a freeway link. Since 37.5 mph is 75% of the speed limit, 0.75 is added to the formula for medium congestion levels. Likewise, 25% is used for the heavy congestion, since the center point is 12.5 mph for the 0 to 25 mph range and 12.5 is 25% of 50 mph. Eqns (2) and (3) represent the EV free flow speed for medium and heavy traffic conditions respectively, allowing for the EV to maintain a 10-mph speed advantage over the normal speeds of vehicles.

$$Medium\ Traffic\ Response\ Time = \frac{L}{SL * 0.75 + 10mph} x\ 60 + 1.2 \quad (2)$$

$$Heavy\ Traffic\ Response\ Time = \frac{L}{SL * 0.25 + 10mph} x\ 60 + 1.2 \quad (3)$$

Signalized intersections also delay EVs during response. Red signals require the vehicles to come to a complete stop before proceeding through, while at green signals the emergency vehicle may continue with minimal disruption. Eqn 4 shows the estimated amount of time it takes to clear a signalized intersection during a green phase and eqn 5 shows the extra time when a red phase is incurred. EVs must begin to decelerate before the intersection and travel 10 mph through a 250-ft approach, coming to a stop at the intersection. Entering a red signalized intersection takes roughly 22 seconds (17 seconds for deceleration and 5 seconds to come to a complete stop). The acceleration time to reach the desired vehicle speed from eqns 1 through 3 takes eight seconds for ambulances. For green phases, simply traveling at 10 mph through the intersection (unless making a turn) will take 17 seconds. The constants for red and green phases are added to the equation, and a key input is the amount of green time for each roadway. Green time varies for every intersection, but for this study, major roadways are assumed to receive 67% of green time due to the typical volume/capacity ratio, while the minor roadways receive the remaining 33%. Although the green ratio is shown, the signal is not necessarily green the entire time. Red and yellow time are factored into each signal with a three-minute cycle length, indicating 62.5% green for major roadways, 37.5% red for major roadways, 29.2% green for minor roadways, and 60.8% red for minor roadways.

$$Green\ Phase\ Intersection\ Time = N * (Green\ \%) * \frac{Clearance\ Time}{60\ sec/min} \quad (4)$$

$$Red\ Phase\ Intersection\ Time = N * (Red\ \%) * \frac{Clearance\ Time}{60\ sec/min} \quad (5)$$

where N = number of intersections per functional class of roadway.

The NEMSIS data, LFUCG data, and HIS data for Fayette County were used to compute estimates of CAV benefit. Due to the unknown starting points for some runs, an assumption is made that every call originates at the nearest fire station housing EMS. Once the distances are calculated, the average response times are computed. Knowing the average run time and time to en-route, ddEMSCAT compares the calculated run time to the average run time, which can then be used to estimate the number of lives saved through time reductions.

Depicting the distances covered by EMS by facility types is challenging because calls occur everywhere. In ddEMSCAT, the HIS from the KYTC is used to estimate EMS mileage on each facility type. HIS shows how many miles of each facility are available in each county. With the mileage of each facility, it is possible to compute the percentage of roadway type each response would require. In Fayette County, only 10% of roadways are freeways, while their VMT is much higher than 10%. With the percent of mileage and speed limit for each facility, the free flow speed can be calculated for EVs.

The time of calls is important for calculating run time. When congestion is light, the EV will obviously have a faster response than calls during heavy congestion. This portion of ddEMSCAT is the last data-driven component. Knowing the percent of each congestion level, the EV response time is calculated with 0% CAV market penetration and compared to the actual (current) average run time from Fayette County. Since market penetration is speculative, the user can from 0% to 100% the market penetration of both autonomous vehicles and connected technology. As only higher levels of automation have the ability to respond to EVs without human interaction, only levels 3.5 to 5 have a user market penetration input cell.

Once all the data and speculative cells are filled, the time savings calculator runs through the average response distance, roadway facilities data, traffic congestion factors, market penetration, and finally the travel time equations to calculate the travel time without CAVs and time with CAVs, finally showing the time savings. The calculations for time savings are indicated in eqns (6)–(10)

$$Travel\ Time\ (TT)\ @\ FFS = Facility\ \% * \frac{Distance}{SL + 10} * 60, \quad (6)$$

$$\%\ TT\ Savings = \left(1 - \frac{TT\ @\ FFS}{TF}\right) * MP, \quad (7)$$

where MP = % market penetration and TF = travel time with traffic.

$$Calculated\ Travel\ Time = (1 - \%\ TT\ Savings) * TT\ @\ FFS. \quad (8)$$

Total travel time for runs can be computed as shown in eqn (8).

$$Total\ Travel\ Time\ = \sum TF * \%\ of\ Mileage, \quad (9)$$

$$Time\ Saved = Travel\ Time\ Today - Travel\ Time\ with\ CAVs. \quad (10)$$

The next component shown in Fig. 2 related to patients. The number of patients is determined by area and time period. The data for this study was limited to Fayette County, Kentucky for January, 2019. Within that month, there were 3,995 EMS runs. The patient cells are populated with 15% non-cardiac arrest, life threatening and 9.7% cardiac arrest patients. The remaining 75% include other calls, still seeking hospital care, yet the injuries or illnesses are not considered to be as time sensitive. Eqn (11) shows the procedure for estimating the number of time sensitive patients. The numbers show the minimum of number of lives saved based on the 1–2% mortality rate for every untreated minute of non-cardiac, life threatening patients and a 7–10% mortality rate for every untreated minute of cardiac arrest patients.

$$TSP = P(CA\% + OTS\%), \quad (11)$$

where TSP = time sensitive patients; P = number of patients; CA% = cardiac arrest percentage; and OTS% = other time sensitive percentage.

The final portion of Fig. 2 shows the link between travel time and patient data. With the maximum, minimum, and average mortality rate calculated for both cardiac arrest patients and life threatening, non-cardiac arrest patients, the total number of lives saved is bolded in the gray box, displayed in Fig. 3. The number of lives saved is calculated using the mortality percentage at the response time multiplied by the number of life-threatened patients and the new travel time, shown in eqn (12)

$$Mortality = TT * mortality\ \% * Number\ of\ Time\ Sensitive\ Patients, \quad (12)$$

where TT = total travel time.

ddEMSCAT

Patients		Response Information			Roadway Facilities (Fayette)				Percent of Calls		Market Penetration		Time Savings	
Number of Patients	3995		Average	Max	Facility Type	Mile-age	%	Speed Limit (MPH)	% During Light Traffic	45%	% AV of Level 0 to Level 3 (Does not contribute to EMS)	0%	Travel Time (Calcu-lated)	4.40
		Distance to Address (Miles)	0.91	6.5	Freeway	69.4	10%	70			**% AV of Level 3.5 to Level 5**	**100%**		
% Life Threatening (Non-Cardiac Arrest)	15.0%	Time fr. Call to On-Scene (Mins)	9.20	66.2	% of 5/7-Lane CTWLTL	43.3	6%	40	% During Medium Traffic	39%	**% V2V**	**100%**	Travel Time with CAVs	1.18
% of Cardiac Arrests	9.7%	Time To En Route (Mins)	4.83	37.2	4-Lane with Turn Lane at Intersection	96.2	14%	45			**% V2I**	**100%**		
Lives Saved due to CAVs Min	**189**	Run Time (Mins)	4.37	48.4	2-Lane with Turn Lane at Intersection	367.9	53%	35	% During Heavy Traffic	16%	Total CAV Contributing	100%	Time Savings	3.22
Avg.	**243**	Travel Time To Hospital (Mins)	8.80	37.2	2-Lane without Turn Lanes	115.2	17%	25						
Max	**296**	Total Time	25.13	138.4										

KEY
User Inputs
Data Driven
Calculated
Calculated

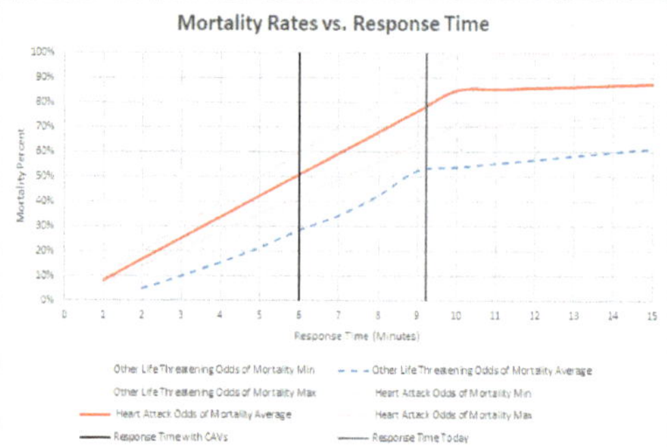

Figure 3: ddEMSCAT interface.

The framework of ddEMSCAT is indicated by the equations shown above. The tool was designed to be straightforward and user-friendly. Fig. 3 shows the ddEMSCAT user interface. User inputs are bold with a white background. The data driven components have a white background and regular text, the calculated portion has a gray background, and the calculated section with number of lives saved has a gray background and bolded text.

The first category is for patient data. The user has the option to place in the number of patients and the percentage of cardiac arrest or percentage of non-cardiac arrest life threatening patients. The numbers in the tool are data-driven from Fayette County. The bottom portion of the patients category shows the minimum, average, and maximum number of lives saved due to the market penetration of CAVs based on the mortality rates and time saved.

The second portion includes response information. The average category shows the average distance traveled to a call, response time, run time, travel time to hospital, and total time. The time to en-route cannot be manipulated by CAVs, but everything else can be reduced by the market penetration of CAVs. The minimums for all the calls are not shown because the category is filled with zeros, as the EMS units were already on scene. An example of an EMS unit being on scene immediately is for an event with large crowds where EMS units are stationed at the event for precaution.

The third portion of ddEMSCAT includes roadway facilities. Since the VMT was not shown for EMS units, the mileage of roadways is shown instead. The number of miles of each roadway is shown with the appropriate speed limit for each roadway type. All the data in this section can be manipulated for different municipalities, but this section includes data from Fayette County.

The fourth portion of ddEMSCAT is the percent of calls. The percent of calls during different levels of traffic congestion was found from Google Maps data for each day of the week in Fayette County. The user as the ability to change the traffic congestion percentage for different municipalities, which could then show the time savings for different congestion levels.

The fifth portion of ddEMSCAT is more speculative. Market penetration of future CAVs is unknown as of today. Thus, the user as the option of changing the percentage of AVs on the roadways, as well as the connected technology percentage. V2V and V2I each contribute 25% of the market penetration, while AVs contribute 50%, eliminating human error.

The final section presents the time savings. All of the user inputs are calculated and shown to calculate the response time of EVs with human drivers and the response time with CAVs. By subtracting the travel time with CAVs from the time without CAVs, the total time saved is shown. The time saved is then added to the average time to en-route to find the total response time. The total response time is placed into the final equation to find the total number of lives saved.

ddEMSCAT users can fill in their own data and speculations to estimate the time savings with CAVs, as well as the number of patients potentially saved. Thus, the contribution of the study is both a framework for analysis and an example using as much real-world data as can be expected in any local community. Of course, the results can be scaled up to regional, state or national levels given appropriate data and assumptions.

3 RESULTS

As of today, 45% of calls are completed within the five-minute critical time, while 41% exceed the eight-minute critical time. The study indicates that for Lexington, a savings of 17% in response time is possible with a CAV market penetration of 35%. For 50% penetration of connected and 50% autonomous vehicles, the potential time savings could attain 68% in decreased run time, approximately 3.2 minutes in Lexington. This could result in an average response time of 6 minutes, still higher than the 5-minute critical time but lower than the 8-minute critical time. Fig. 4 gives provides a visualization of call times with market penetrations of 0%, 50%, and 100%. As can be seen, as the market penetration increases, the number of calls over five minutes and eight minutes decreases within the city.

Travel time savings are important, but the goal is to save lives. To compute the effect of the savings on mortality, the bottom section of Fig. 3 shows the estimated mortality rates over time with the present response times as compared to response times with CAVs. The figure depicts a full market penetration of CAVs which brings the response time down from 9.2 to 6 minutes. Note that the average mortality rate of cardiac arrest patients at 9.2 minutes is 78%, while the average mortality rate for other life-threatening patients is 53%. At 6 minutes, the mortality rates for the two cases decrease to 51% for cardiac arrests and 29% for other life-threatening conditions. While these percentages are speculative maximums, the ddEMSCAT framework allows the user to input data into the tool to create a forecast for different cases and areas.

Figure 4: Number and location of ambulance calls over critical response times with different market penetration levels of CAVs.

4 ITALIAN IMPLICATIONS

On the face, this work would seem to be equally applicable in Italy or any industrialized country's urban areas. Key is the availability of data to populate the inputs to ddEMSCAT for the computation of potential savings. While we were not able to obtain Italian data in time for publication of this paper, there are several differences between USA and Italian EMS services that could (a) make results more or less accurate, and (b) require additional factors in ddEMSCAT. For example, Italian EMS follows the Franco-German Emergency Medical Services System (FGS), where the main motto is to "Stay and Treat", whereas in the USA, the Anglo-American emergency Medical Services Systems (AAS) is followed, which aims at "Transport and Treat" [16].

In the USA and in England, the crews in an ambulance serve to stabilize the patients before they can be transported to a nearby facility for further treatment. But in Europe, and Italy in our case, ambulances are so well equipped that they are prepared for treating a critical patient on site. But to do that, the crews and staff need to be very well trained [17]. This difference between Italian and US services does not directly affect response time, but there are likely indirect effects. For example, as the Italian ambulance staffed by a physician is perhaps better suited to enable a better outcome for an individual patient, if the number of units per capita is fewer, dispatch areas could be larger. However, due to the higher population density of Italian cities and regions as compared to the US, the areas may in fact be smaller.

A study comparing EMS services in the US and EU reports Genoa, Italy has less than a quarter or the EMS Transports as Richmond, Virginia, USA. Obviously, if this ratio is typical and representative of life-threatening EMS calls, the benefits of reduced response times as calculated by ddEMSCAT would be much smaller in Italy. However, the same study reports that Genoa has 25% more attempted EMS cardiac arrest resuscitations [18]. The population of the Genoa area is approximately 20% greater than that of the Richmond MSA, suggesting that the ratio of life-threatening calls between the two cities is probably closer to parity. These and other differences require further study but could be accounted for in the ddEMSCAT framework.

5 CONCLUSIONS

The results of this work indicate the potential for CAVs reducing emergency response times in urban areas by following rules, setting examples, and ultimately leading to the introduction of the virtual emergency lanes, especially near intersections. This work addressed several of the theoretical boundaries using current data and attempting the future forecast with the potential of CAVs. The tool developed for estimating these benefits could be enhanced with the consideration of binning response times. Longer response times would likely see the most dramatic change under CAV technology scenarios, though critical improvements to key response times may have the most profound effect on lives saved. For instance, most time-sensitive emergency conditions are sensitive to the five-minute and eight-minute response marks.

By assuming different levels of market penetration and understanding the ways EVs and CAVs could potentially interact, we can see that CAVs can reduce the average emergency response time by over three minutes. In our case study of Lexington, Kentucky USA, this could result in a reduction in cardiac arrest mortality as well as non-cardiac arrest and life-threatening mortality rate by 35%.

This study has limitations typical in any forecast of future technology. Lack of data necessitates many assumptions, including time required for proper care, locations of calls, and simulations. This report and ddEMSCAT lay out a framework for investigating emergency response times facilitating the pursuit of additional needed research into off-system impacts of CAVs on public health.

ACKNOWLEDGEMENTS

The authors thank Joe Nugent and Todd Samuelson from the Lexington Fire Department. We are also thankful to A.M. Hasibul Islam and Riana Tanzen for reviewing literature. This research was partially sponsored by Dwight D. Eisenhower Transportation Fellowship Program.

REFERENCES

[1] Heron, M., Deaths: Leading causes for 2016. *National Vital Statistics Reports*, **67**(6), 2018.

[2] Pell, J.P. et al., Effect of reducing ambulance response times on deaths from out of hospital cardiac arrest. *BMJ*, **322**(7299), pp. 1385–1388, 2001.

[3] LFUCG, Geographic information services. www.lexingtonky.gov/departments/geographic-information-services. Accessed on: 8 Mar. 2019.

[4] Bureau of Emergency Medical Services, *2018 Mid-Year Data Report*, 2018.

[5] Lum, G.M.L., *Impact of Automatic Vehicle Locators on Incident Response Times in the Honolulu Fire*, 2008.

[6] RapidSOS, Outcomes: Quantifying the impact of emergency response times. https://cdn2.hubspot.net/hubfs/549701/Documents/RapidSOS_Outcomes_White_Paper_-_2015_4.pdf. Accessed on: 23 Aug. 2019.

[7] Link, M.S. et al., *Part 6: Electrical Therapies: Automated External Defibrillators, Defibrillation, Cardioversion, and Pacing*, 2010.

[8] Blackwell, T.H. & Kaufman, J.S., Response time effectiveness: comparison of response time and survival in an urban emergency medical services system. *Academic Emergency Medicine*, **9**(4), pp. 288–295, 2002.

[9] Blanchard, I.E. et al., Emergency medical services response time and mortality in an urban setting. *Prehospital Emergency Care*, **16**(1), pp. 142–151, 2012.

[10] Jordan, C.A., Cetin, M. & Robinson, R.M., Path clearance for emergency vehicles through the use of vehicle-to-vehicle communication. *Transportation Research Record: Journal of the Transportation Research Board*, **2381**(1), pp. 45–53, 2013.
[11] World Health Organization, *Emergency Medical Services Systems in the European Union, Data Book*, 2008.
[12] NEMSIS, Annual implementation meeting. https://nemsis.org/. Accessed on: 8 Mar. 2019.
[13] KYTC, Planning highway information (HIS database). https://transportation.ky.gov/Planning/Pages/Roadway-Information-and-Data.aspx. Accessed on: 31 Mar. 2019.
[14] Google Maps, https://maps.google.com/. Accessed on: 31 Mar. 2019.
[15] Biersdorfer, J.D., Q&A: Color-coded traffic in Google Maps. https://gadgetwise.blogs.nytimes.com/2010/08/16/q-a-color-coded-traffic-in-google-maps/. Accessed on: 8 Apr. 2019.
[16] Dick, W., Anglo-American vs. Franco-German emergency medical services system. *Prehospital and Disaster Medicine*, **18**(1), pp. 29–37, 2003.
[17] Mitchell, J., Physician aids in overhaul of Tuscan emergency medical system. *Tufts Journal*, 2005.
[18] Overton, J., Benchmarking EMS in Europe. 7th Congress of the European Resuscitation Council. www.geographie.uni-muenchen.de/eed/assets/presentations/EED_Benchmarking.pdf. Accessed on: 21 Dec. 2005.

FIRE INVESTIGATION IN AN ITALIAN WASTE TREATMENT PLANT: LESSONS LEARNED AND FUTURE DEVELOPMENT

GIOVANNI ROMANO & ALFREDO ROMANO
RSM Romano Safety Management STP, Osio Sotto (BG), Italy

ABSTRACT

In June 2017, in the north of Italy, a serious fire occurred in a waste treatment plant and involved the entire warehouse. Romano Safety Management (RSM) was charged with reconstructing the evolution of the fire, in order to establish whether the sequence of events leading up to the accident could be compatible with developing times of the accident. Documental evidence has collected in the accident place, as well as proofs of operators and rescuers who provided support on-site. The fire, the structure and materials composing the warehouse were rebuilt using "Fire Dynamic Simulator" (National Institute of Standard and Technology) and "PyroSim" (Thunderhead Engineering). RSM carried out a three-dimensional simulation of the fire, assuming that the origin of the fire was provoked by an accumulation of crushed waste of mattresses and considering interferences with the other parts of the warehouse, where different materials (wood, rubber, etc.) were stocked. According to the results, considering the ignition temperature of products and the presence of concrete separation walls, RSM concluded that the simulation of fire development was congruent and compatible with the development times of the fire. At the end, it was demonstrated that there were no active or passive fire-prevention protections inside the warehouse; these protections, jointly with a better operative management of the warehouse, could have avoided the fire development. In the last few years, many similar fires occurred in Italy and, for that reason, the Italian Ministry of Environment recently issued the first version of a "Guideline" according to risk management. These accidents and the results of the analyses will probably lead to a review of the fire-prevention design criteria of these plants.

Keywords: fire investigation, waste treatment plant, fire safety engineering, performance approach, fire risk analysis.

1 INTRODUCTION

The purpose of this article is to share the experience gained by RSM (Romano Safety Management) in "fire investigation". This is a fire that developed inside an Italian industrial plant treating various types of waste. Some details are omitted because a judicial inquiry is still under way. RSM had been commissioned by its Client to proceed with a simulated "reconstruction" of the evolution of the fire. In particular, the propagation speed, intensity and persistence of the flames were investigated, in order to verify whether the fire was compatible with the materials present at the time of the accident and to verify the ignition condition of combustible material placed in the warehouse.

The reconstruction of the simulation model, and in particular the materials inside the compartments, was obtained thanks to the elements provided by the Client and the verification of the evolution of the fire through interviews with staff, paper documentation and photographs.

The fire simulation was performed through the calculation models called "Fire Dynamics Simulator" and the "PyroSim" interface. The definition of the fire scenario assumes a fundamental role in the characterization of the fire itself and its results; in the case under examination, RSM predefined it on the basis of the elements collected and not through a risk analysis that is usually carried out upstream the identification of the scenario itself.

www.witpress.com, ISSN 1743-3509 (on-line)
doi:10.2495/SAFE190111

The purpose of the analysis was to clarify some aspects of the development of the fire, in order to evaluate appropriate improvement actions to prevent similar accidents, and in particular:

a) if the fire in the warehouse was congruent in terms of propagation speed and intensity;
b) if the fire in the outside container of combustible material was compatible with the development of the fire inside the warehouse that would generate temperatures of degradation of the product contained therein and consequent ignition;
c) once confirmed the non-compatibility, what could be the probable causes that led to the fire.

2 SUMMARY OF THE ACCIDENT AND IDENTIFICATION OF THE CRITICAL PARAMETERS OF REFERENCE

The fire developed from a combination of products deriving from the treatment of combustible materials (it has to be underlined again that some details are deliberately omitted from the present). For the purposes of the present, the potential causes of ignition have not been analysed, but the sequences of ignition of the other materials in the warehouse have been analysed. The critical points of the evolution are the following:

Time 0:	the operators, which were present, detected a beginning of fire during the normal treatment operation.
Time 180 s:	an operator tried to "isolate" the inflamed material from the one not yet affected by the fire, by taking the inflamed part of the pile and bringing it outside, the operation partially succeeded.
Time 300 s:	the fire was extended to the compartment near the pile of combustible material and, in the meantime, the internal emergency team was operative to prevent the fire as much as possible.
Time 300–600 s:	the vehicles inside the warehouse were moved outside from the side exit at about 40 metres.
Time 600 s:	In the time between 300 and 600 seconds, the fire extended to the other sectors where there were other combustible materials.
Time 15–20 minutes:	all the piles of rubber products were on fire and the warehouse was completely full of smoke.

3 CALCULATION MODELS USED

The fire simulations described in the previous paragraph were carried out by using the Fire Dynamics Simulator calculation model (known as "FDS", acronym used below) by the National Institute of Standards and Technology (NIST, USA).

FDS is a Computational Fluid Dynamics calculation model ("CFD"), specifically developed for fires, that uses LES [1] techniques ("Large Eddy Simulation"). This model permits us to foresee the heat conditions resulting from a fire in a compartment [2], [3].

A CFD calculation model requires that the compartment or building of interest (intended as a series of compartments) be subdivided into control volumes (computational cells) in the form of a parallelepiped or, better, cubic. The CFD calculation model estimates the density, speed, temperature, pressure and concentration of the chemical species in each cell.

Based on the laws of conservation of mass, moment, species and energy, the model outlines the generation and movement of fire gases. FDS employs the material characteristics of the elements of the survey domain, in order to estimate the growth and spread of the fire. A complete description of the model can be found in the documentation annexed to it (see previous notes).

The ability of the FDS model to predict accurately the temperature and speed of the gases developed by the fire was previously evaluated by the NIST by conducting appropriate experiments, both on a reduced scale and in a real scale and the subsequent analysis of the data from the field. For relatively simple flows caused by the fire, such as in the case of naked flames and flows through passages, doors and windows, the FDS predictions are compatible with the experimental uncertainty of the values measured in the experiments conducted over the years by NIST (since 1998).

In large-scale test fires, temperature predictions through the use of FDS showed values characterized by a difference of 15% compared to the actual measured temperatures, while the Heat Release Rates predicted by a difference of 20% of the measures taken.

4 GEOMETRIC RECONSTRUCTION AND CHARACTERISTICS OF PRODUCTS IN THE WAREHOUSE

The geometric reconstruction of the building, which is necessary for the modelling of the fire scenario, was carried out by using the PyroSim 2017 software. We proceeded to a three-dimensional survey with laser scanning technology and to a subsequent processing of the point cloud through 3D processing software. Fig. 1(a) shows a screenshot of the point cloud, and Fig. 1(b) the three-dimensional model subsequently imported into PyroSim.

(a) (b)

Figure 1: (a) Point Cloud from laser scanning; (b) 3D model in PyroSim. Inside the building, there were no fire-fighting compartments.

The warehouse was equipped with a bird mesh screen that, in the simulation, was considered as a propylene net with density of 1 kg/m³ and melting temperature of 160°C.

The HRRPUA parameter (Heat Release Rate Per Unit Area) represents the heat flow developed by combustion per unit area.

Table 1 shows the main characteristics of the construction materials.

Table 1: Characteristics of construction materials.

Material	Density (kg/m³)	Heat conductivity (W/m K)
New Jersey barriers	2280	1.8
Sandwich panels	40	0.02
Polycarbonate glasses	1380	0.2

In the attic, there was also a polypropylene mesh. Table 2 below shows the typical data of the materials used for modelling the fire scenario.

Table 2: Typical data of materials used for the simulation.

Material	Density (kg/m^3)	Auto-ignition temperature (°C)	HRRPUA (heat release rate per unit area) (kW/m^2)
Material 1	50	250	500
Material 2	640	450	200
Material 3	640	400	200
Material 4	900	350	300
Rubber	1400	400	500

5 IDENTIFICATION OF FIRE SCENARIO

The study was carried out by a three-dimensional fire simulation, assuming that the fire source originated from a pile of combustible waste, placed near the entrance of the warehouse, formed after crushing operations. According to the reconstruction carried out, the initial fire ignition started from the burning of the crushed pile.

The simulation was carried out without considering any fire mitigation (it should be noted that, after about 300 seconds, the internal emergency team intervened to start the first cooling operations, waiting for the Fire Brigade arrival).

6 FIRE SIMULATION

The result of the three-dimensional simulation with the FDS/PYROSIM calculation model consists of a video that allows checking the development of the fire according to time and temperature profiles on the walls of the warehouse and on the virtual plans introduced during the input phase, in particular on the caissons and on the accumulation of the products. The analysis of the video highlighted the following interesting points, to which we extracted "still images" with related comments. The 33 s still image (Fig. 2) shows that the accumulation of combustible material is partially ignited but it has not yet initiated and propagated to the compartment next the combustible material.

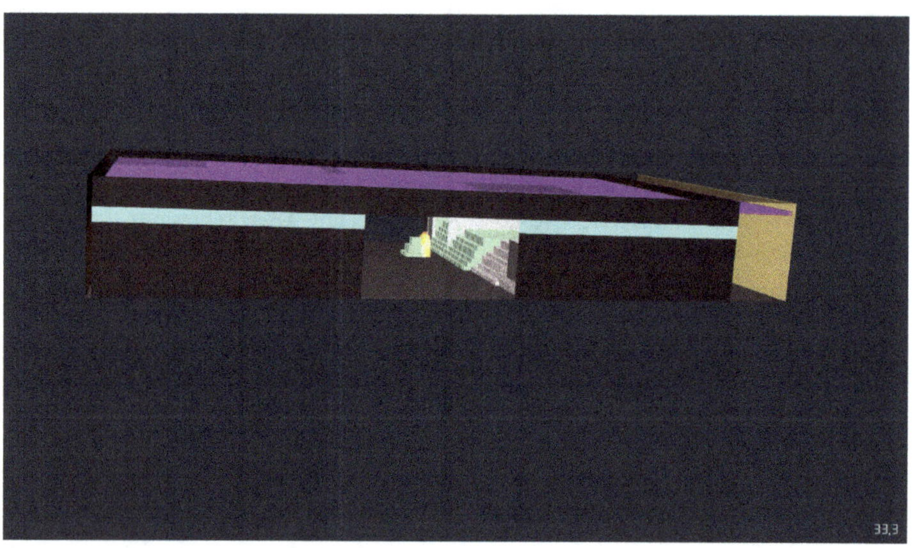

Figure 2: Accumulation of fire crushed material (Time 33 s).

The accumulation of crushed material, from which the fire began, ignited the compartment next to the combustible material on the wall side of the warehouse: Fig. 3 shows the initial instants of the fire, in which the temperatures are still quite basic (compare the colour of the wall and the temperature scales).

Fig. 4 shows that the fire is propagating to the compartment next to combustible material after about 147" and, therefore, is in accordance with what was reported in the description of the accidents. Fig. 5 shows the fire involving the whole warehouse.

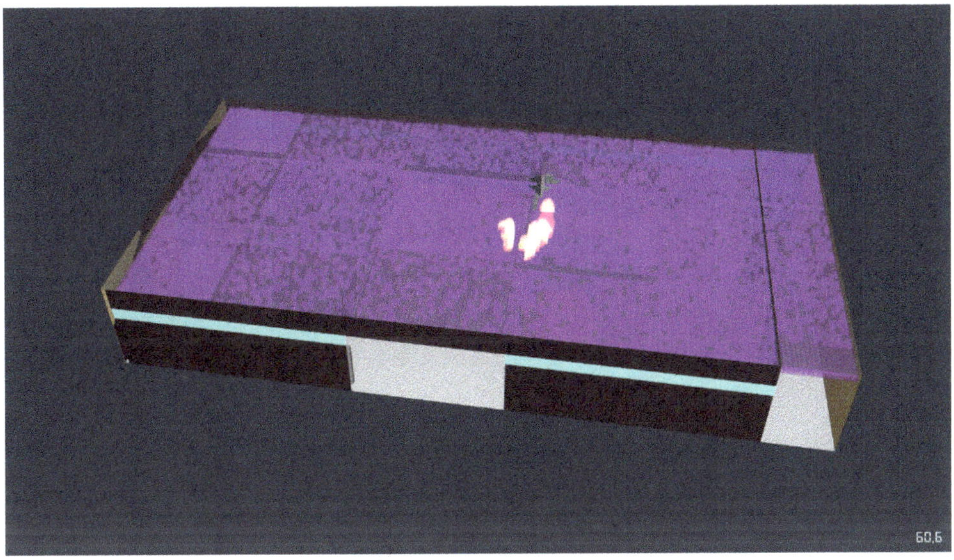

Figure 3: Accumulation of fire combustible material (Time 60 s).

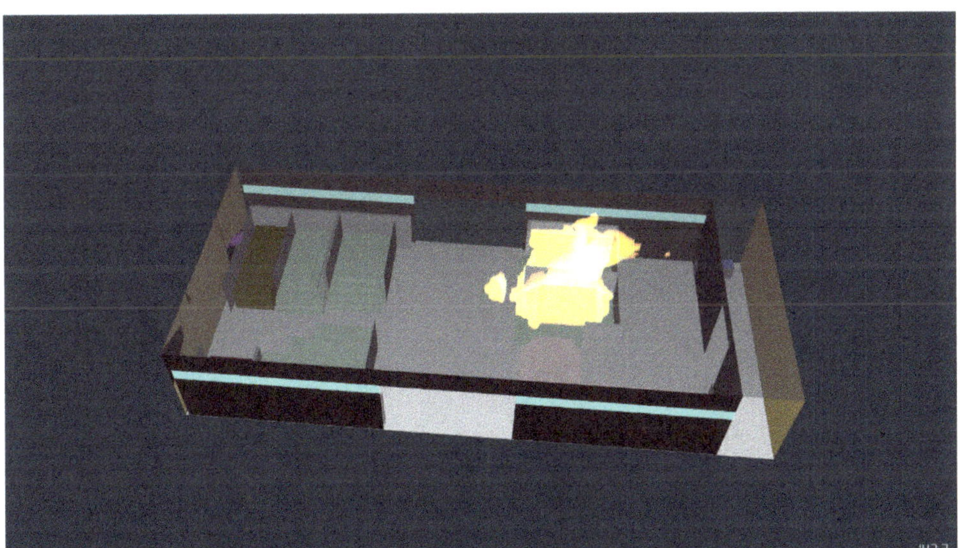

Figure 4: Propagation of fire to compartments (Time 147 s).

The development of the fire is congruent with the available elements. The temperature profile detected with a virtual plane on the tire caisson leads to low temperature values not able to develop the degradation of the rubber, in order to facilitate ignition.

The fire (Fig. 6) continues even if the piles of crushed material to the next compartments begin to decrease in consistency and the flames are lowered with consequent reduction of the temperatures; it should also be noted that in the compartment 2 (completely wood) there is no ignition.

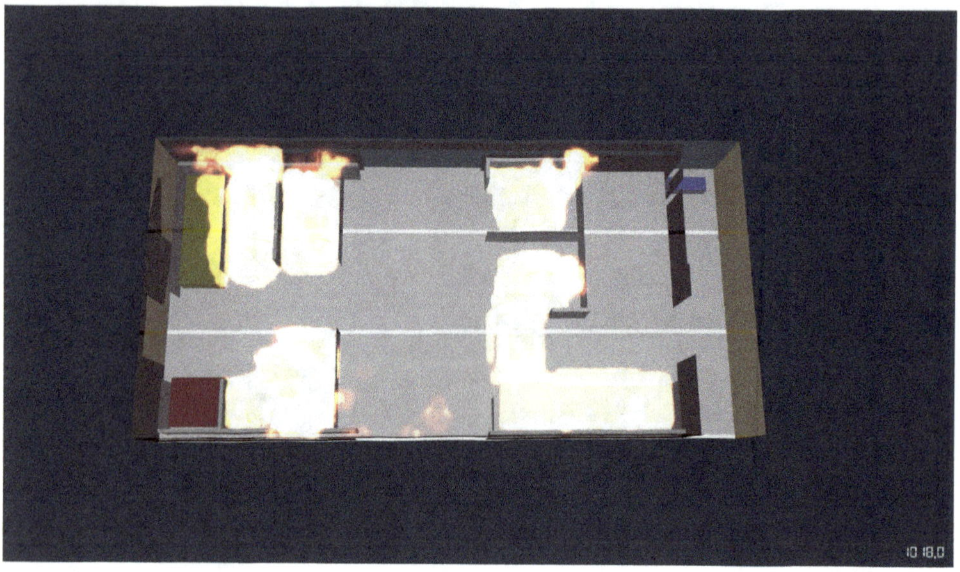

Figure 5: Fire involving the entire warehouse (time 1000 s).

Figure 6: Temperature profile (time 1000 s).

7 RESULTS ACHIEVED

The simulation demonstrated that the development of the fire in the outer caisson of combustible material was not due to the presence of flammable products that could be a source of possible secondary ignition. The simulation also showed that the development of the fire inside the caisson was not compatible with the temperature reached inside the warehouse. This statement is further strengthened by the conservative assumptions upstream of the simulation, which aggravated the incidental scenario with a pile of additional fuel product close to the combustible material, from which the ignition began and it did not take into account the mitigation measures carried out by the internal emergency team and the subsequent intervention of the Fire Brigade.

The videos of the three-dimensional simulation of the fire and the sequence of photos above confirm that inside the caisson, a temperature causing a decomposition of the rubber and leading to hydrocarbon vapours, which usually release from this type of product, was not reached.

Actually, however, it was found that the fire was there and, then, it is necessary to proceed with some reflections, in order to understand and reconstruct what happened and why the results of the simulation are quite congruent with the development of the fire within the warehouse and, instead, it seems not considering the caisson placed outside.

The possible hypotheses that may have led to this lack of characterization can be attributed, in our opinion, to the state of the areas on the caisson-side that, in the modelling, was considered "empty" of materials from the compartments next to the origin of the ignition up to the wall bordering the caisson and the area surrounding the caisson itself.

8 FURTHER DEVELOPMENT

In the last few years, many similar fires occurred in Italy and, for that reason, the Italian Ministry of Environment has recently issued a first version of a "Guideline" according to risk management.

These accidents and the results of the analyses will probably lead to a review of fire-prevention design criteria of these plants, such as firefighting plant (fixed or mobile), gas or fire detection system and a critical review of Emergency Internal Plan.

During the analysis it was found that, at a design level, some measures could be taken, such as for example passive protection or the insertion of a dedicated sprinkler system, even limited to a portion of the warehouse.

The fire and/or smoke detection could be considered even if the intervention of the team was timely and the high fire risk processes are carried out during the personnel supervision.

REFERENCES

[1] McGrattan, K. et al., Fire dynamics simulator (version 5) technical reference guide. NIST Special Publication 1018-5NIST in cooperation with VTT Technical Research Centre of Finland, National Institute of Standards and Technology, US Department of Commerce, 2008.

[2] McGrattan, K.B., Hamins, A. & Stroup, D., *Sprinkler, Smoke & Heat Vent, Draft Curtain Interaction – Large Scale Experiments and Model Development*, NISTIR 6196-1, National Institute of Standards and Technology: Gaithersburg, 1998.

[3] McGrattan, K.B., Baum, H., Rehm, R. & Large, G.E., Simulations of smoke movement. *Fire Safety Journal*, **30**, pp. 161–117, 1998.

RISK MANAGEMENT PRACTICES IN PROJECT ACTIVITIES AT NORTH KUWAIT OIL FIELDS

AHMED M. QABAZARD
Health, Safety & Environment (North Kuwait) Team, Kuwait Oil Company, Kuwait

ABSTRACT

Risk management is a concept which has become very popular in a number of business organizations. Many companies often establish a risk management procedure in their projects for improving the performance and increasing profits. The project construction industry is highly risk prone, with complex and dynamic environments which create an atmosphere of high uncertainties and potential hazards. The project sector is vulnerable to various technical, socio-political and business risks including HSE (health, safety and environment). The track records to cope with these risks have not been very good in the construction industry. As a result, the people working in the project industries bear various failures, such as failure to abide by quality and operational requirements, exceeding cost and uncertain delays in project completion including HSE obligations. The risk based approach or risk management is an important process of project activities which consist of the identification of risks, appropriate assessment, responses with a suitable method for handling risks and then developing strategies to mitigate those risks in compliance with the regulatory objectives and the company's risk management expectations. This paper covers the concepts of HSE risk management through various risk evaluations, project challenges and effective risk mitigation/reduction practices being followed at Kuwait Oil Company (KOC) – North Kuwait (NK) oil fields – for the one stop solution for all types of hazards and enhancing HSE compliance during any construction project lifecycle. The risk based approach consists of both proactive and reactive control and corrective action that could be applied according to the nature, severity and immediacy of the risk posed, which enables one to be flexible, adaptive and focused to the project HSE demands and ongoing changes in the project plan without affecting contractual HSE obligation. With the above approach, there is a visible improvement in HSE compliance at contractors' worksites at NK oil fields, which signifies the company's HSE commitment in its business process.
Keywords: risk management, project risk.

1 INTRODUCTION

The role of contractors in industrial development, especially in oils sectors, are very vital. Industrial sector cannot develop with the current growing demand without the assistance of proper contracting agencies. The contractors are key players in the sustainable oil sector projects from predesign and design, through construction and project closeout. They also provide immense support in operation and maintenance of our production facilities. However, in this context a question always arises in our mind that, how do the project/ construction risks are managed effectively? There are high potential risks involved in the project activities including key challenges and their implications.

In order to address the above challenges, we follow some of the monitoring practices as a part of project risk management to ensure HSEMS (health, safety and environment management system) compliance [1] at contractors' worksites through regular site inspections, site verification visits, mobile equipment integrity surveys, audits, on- site awareness, campaigns, etc. These monitoring practices are not only to ensure site compliance, but also provide the opportunities for continual risk improvements.

2 NECESSITY

Globally it is acknowledged that an improved HSE compliance implementation through risk-based approach has direct and strategic bearings on functional and operating project HSE

www.witpress.com, ISSN 1743-3509 (on-line)
doi:10.2495/SAFE190121

performance standards [2] of various contractors. Project accomplishments involve various intended activities of pre-design, design, review, construction, testing, inspection, commissioning, stabilization process, etc. to be carried out by multiple agencies within the specified period. Sometime this involves surprises (unplanned activities) due to various reasons. Because of constraints of time and resources, there is a possibility of bypassing the obligatory/regulatory HSE compliance leading to enhancement of risk level to undesired/ unacceptable level. Hence, the established and effective risk management practices will ensure proper identification of risk, its' assessment and suitable control/mitigation measures in the method for handling risks and then develop strategies to mitigate those risks in compliance project activities.

3 OBJECTIVE

The purpose of this risk management practice is to evaluate the HSE performance of North Kuwait-based contractors [2] by reviewing their worksite HSE compliance level, policies, statistical performance, adherence to standards, etc. The areas of improvement and positive observations identified through this process are used as a basis for the risk scrutiny and resultant recommendations, in order to improve upon HSEMS procedural compliance [3] at contractors' worksites. The objectives are summarized as follows:

- HSE surveillance (third-eye view) at contractors' worksites;
- Timely HSE support;
- Proper risk assessments of all critical tasks;
- Recording the lessons learnt for future.

The long-term intent of the risk management practices is to build project findings to document best practices, lessons learned, and examples of various documents to benefit KOC and the contractors in terms of benchmarking, evaluation and enhancing HSEMS performances.

4 CHALLENGES OF PROJECT RISK MANAGEMENT

Project construction activities are risk level with complexities, which pose environment of high uncertainties and potential hazards. Each project task is critical and have its own importance and intricacies in handling the activities. Some of the inherent challenges faced in managing HSE compliance during the project execution are as follows:

- Intense competition for available work;
- Multiple critical activities;
- Diverse agencies;
- Diminished work supervision;
- Migrant/new work force in changing environment;
- Management of waste generated out of construction;
- Project schedule goals and targets for safety, quality and cost;
- Multiple agencies with combination of experienced and novice workforce having different age groups poses greater risk and challenge in execution of safe work practices;
- A high number of litigation cases as well as disputes.

5 METHODOLOGY

The above risk management process is unique and different from various conventional methods, which focuses not only on level of implementation but also emphasis on deviations observed during the course of verification. The following steps are involved in the above process.

5.1 Developing a comprehensive analytical tool/framework utilizing various inspection databases

5.1.1 HSE inspections at project worksites

There are comprehensive, planned and periodic inspection schedules prepared for each project sites to provide accurate and objective feedback to management and the adequacy, effectiveness, and efficiency of contractors' HSE programs, compliance and activities compared to applicable requirements. The objectives of the inspection process (Fig. 1) are to evaluate the progress of achieving HSE targets, maintaining regulatory compliance, and effectively utilizing the KOC health safety and environment management system [1] including reporting deficiencies in meeting KOC HSE requirements and commitments. Results of the evaluation of the collected inspection evidence compared with the inspection criteria or the results of an inspection. Findings can indicate conformity or nonconformity with audit criteria, good practices and/or opportunities for improvement.

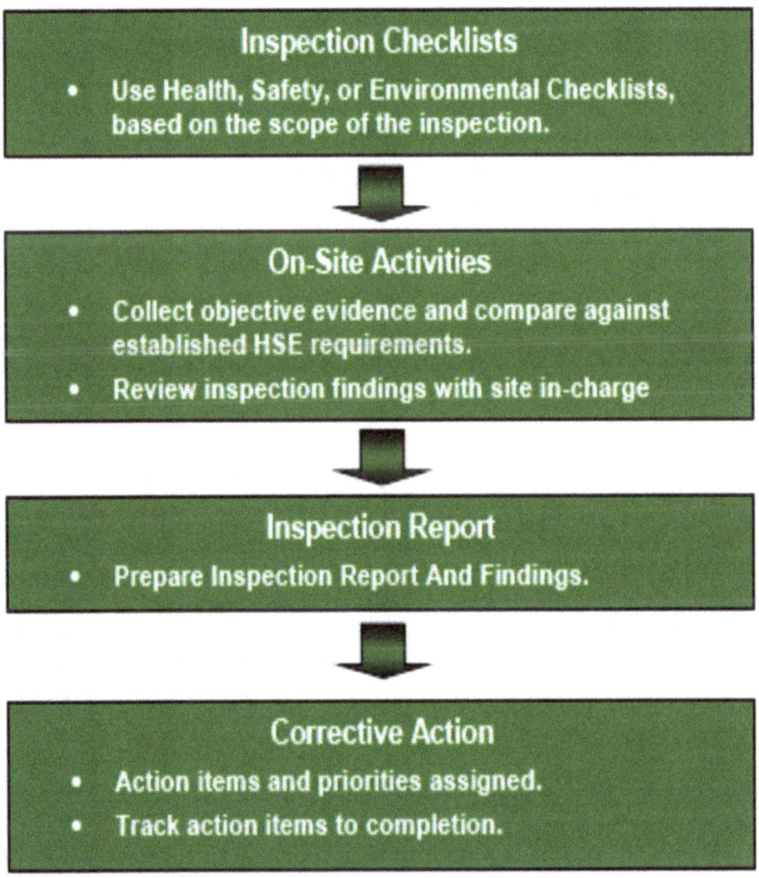

Figure 1: HSE inspection process.

5.1.2 Observations from site verification visits (SVV's)
Site verification visits are the planned visits to the contractors' worksites (typical project worksite as shown in Fig. 2) by either the group or the individual and verify the level of HSE compliance at site and any deviations observed shall be discussed at the visit location/spot and highlighted to the respective controlling teams the project. Usually these SVV are conducted focusing on the HSE themes identified specifically for the month. The management team shall carry out site visit for the specific location indicated in the visit schedule. The main objective of this site visit is to enhance HSE commitment and leadership demonstration at worksite. The observation during site verification visit shall be focused to issues enlisted in the specific checklist as applicable. The observation made by the visitor (management personnel) during such visit shall be reported online through MyHSSE [4].

Figure 2: Typical project worksite.

5.1.3 HSE inspections at project site offices
It is similar process like Inspection at project sites. However, it emphases on the compliance requirements as per KOC standard for portacabins especially on material of construction, electrical installations/fixtures and fire protection arrangements considering the fire hazard potentials.

5.1.4 Other HSE leading and lagging indicator performance from MyHSSE [4] data
Making a HSE program that includes the right elements for assessing HSE performance, like leading and lagging indicators, is a critical step in boosting overall HSE performance. Leading and lagging indicators have a valuable effect on measuring HSE performance. Lagging indicators are reactive in nature. They measure the effectiveness of a safety program after the facts. Typical lagging indicators include the number of incidents, injuries, LTIs (lost time injuries), incident frequency rates, etc. Whereas, leading indicators are proactive in nature. These indicators focus on the future. They target events used to drive and measure activities designed to prevent and control behavior and injuries. Typical leading indicators include the safety training, audit/inspections conducted, etc. Both leading and lagging indicators are relevant to workplace safety, especially at project sites and worth measuring. They present important aspects of an overall HSE management system that drives all to a zero-incident at worksite. All relevant data are entered and logged appropriately in a system-based programme MyHSSE [4] specially design to record and track leading/ lagging indicators.

5.1.5 Tracking of action items in the project risk registers

The main purpose of the preparing the project risk register to identify occupational safety hazards, assess the risks arising from project activities and ensure its proper management in order to comply with legal and regulatory requirements, company HSSE policy [5] and occupation prerequisites. The entire process of preparation/review of risk registers is conducted through series of workshop sessions involving experienced team of multidisciplinary experts from various teams, who brainstorms on each risk aspects pertaining to scope of project work/activity and accordingly ensure that the identified risks are assessed, prioritized and controlled to eliminate the hazards or reduce the risks to acceptable levels. The typical flow-diagram of risk assessment and management process is shown in Fig. 3.

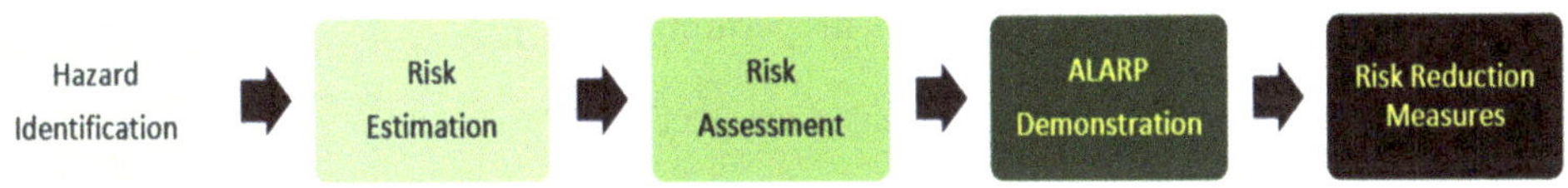

Figure 3: Risk assessment and management process.

5.2 Evaluating HSEMS procedural/standards implementation and compliances based on the risk level

All HSE observations/deviations made at work sites and site offices of all KOC North Kuwait contractors are critically assessed based on the company's risk rating matrix (Fig. 4) as prescribed in KOC HSE management system (HSEMS) procedures [3] and categorized/listed as low (acceptable), medium (tolerable), high (intolerable). The tracking/monitoring of those listed observations are carried out appropriately through a task force team (TFT).

		Severity				
		Negligible (1)	Minor (2)	Moderate (3)	Major (4)	Severe (5)
Likelihood of Occurrence	Very Unlikely (A)	Low (A1)	Low (A2)	Low (A3)	Low (A4)	Medium (A5)
	Unlikely (B)	Low (B1)	Low (B2)	Medium (B3)	Medium (B4)	Medium (B5)
	Possible (C)	Low (C1)	Medium (C2)	Medium (C3)	Medium (C4)	High (C5)
	Likely (D)	Low (D1)	Medium (D2)	Medium (D3)	High (D4)	High (D5)
	Frequently (E)	Medium (E1)	Medium (E2)	High (E3)	High (E4)	High (E5)

Figure 4: Risk rating matrix.

5.3 Identifying and analyzing the positive observations/compliance on KOC HSEMS procedures/standards [3]

Ensuring compliance to HSE procedure/standard and maintaining a safe workplace will reduce the risk of work-related illnesses and injuries, etc. The HSE adherence and good safety practices at worksites are also recorded and provided with equal weightage while comparing the overall HSE performance assessment.

5.4 Identification and analysis of critical risk-based gaps/non-compliance of HSEMS procedures/standards [3]

5.5 Sharing results/contractor performance status with respective project controlling teams to improve contractor performance [2]

5.6 Deliberating on performance status with respective contractors during monthly HSE meetings

5.7 Close follow-up with the respective contractors through their project controlling teams for tracking/pursuing and early closure of deviations/ HSE observations made during the surveillance process/system

6 HSE FOCUS/EMPHASIS

Considering the complexities of the various project activities, other priorities and management expectations during project job execution, more emphasis/importance have been given on the following HSE priority areas.

6.1 Project HSE plan/job execution plans

It is an essential project HSE document that ensures the requirements and expectations of the KOC HSSE policy [5], KOC strategic objectives, regulatory compliance, recommendations of risk insurers, risks and impacts associated with the project, key activities, roles, and responsibilities etc. are addressed adequately and appropriately. This document also explains the relevant project risks are identified and assessed using appropriate tools including mitigation through application of risk management techniques. This document is also reviewed and revised as per the various stages of the project including the project scope.

6.2 Hazard management and communication

The goal of the hazard management and communication is to ensure the cost-effective, integrated actions that will reduce or prevent potential risks to all stakeholders.

6.3 HSE awareness and campaigns to elevate the workers' awareness level

The HSE awareness and campaign is an effective way to spread a safety-related message to employees/workers. It is a structured course of action to achieve a specific goal of shaping the behavior of the intended target audience and providing practical solutions to HSE issues that have been identified.

6.4 HSE surveillance to find the deviations

HSE surveillance is to be more vigilant for safety event detection (third eye view) at various workplaces/worksites revealing HSE deviations/unsafe practices. The objective of the HSE surveillance process is to ensure greater risk reduction.

6.5 Audit and compliance to HSE management system [1]

HSEMS audit is a comprehensive review of company's adherence to HSEMS [1], [3] procedures and standards. Audit reports evaluate the strength and thoroughness of compliance preparations, HSSE (health, safety, security and environment) policy [5].

6.6 HSE reinforcement/facilitation (assistance)

The main objective of above is to ensure provide all necessary HSE support and assistance (as applicable and required) to the workforce working under various risks to mitigate the risk level.

6.7 Support to critical/special project activities

Providing HSE support for critical and special project activities, which need coordinated review and risk assessment by multi-disciplinary team.

7 RISK MANAGEMENT PRACTICES

Project HSE risk management encompasses the assessment and mitigation of project HSE risks. The objective of the above is to assess the risks associated with identified hazards and to develop and implement effective and appropriate mitigations. Project HSE risk management is therefore a key component of the process for both the company and the contractor (service provider). Project HSE risks are conceptually assessed as low, medium, or high (acceptable, tolerable, or intolerable) (Fig. 4). Risks assessed as initially falling in high (intolerable) region are unacceptable under any circumstances. The probability and/or severity of the consequences of the hazards are of such a magnitude, and the damaging potential of the hazard poses such a threat to safety, that immediate mitigation action is required.

Bearing above HSE focal areas (as stated in Section 6) in mind, HSE-NK team has established and following multi-functional proactive measures (as given below) focusing primarily on enhancing the awareness and compliance level at all project worksites ensuring HSE visibility, as a part of risk management practices under accident prevention program at North Kuwait Directorate.

7.1 Project HSE plan

Preparation of appropriate project HSE plan ensuring the requirements and expectations of the KOC HSE policy, KOC strategic objectives, regulatory compliance, recommendations of risk insurers, etc. are all addressed. Also ensuring the potential health, occupational and process safety hazards/environmental aspects and corresponding risks and impacts associated with the project are identified and assessed using appropriate tools and are mitigated through application of risk management techniques. The above plan identifies the key activities, roles, and responsibilities that will meet the HSE goals at the various stages of the project. The same plan also serves as input, at the early stages of the project, for estimation of the time and resources (monetary and human) required for achieving the project HSE objectives.

7.2 Project execution HSE plan and work method statements

Subsequently, respective contractors develop their project execution HSE plan that contains detailed HSE plan and programs that describe how the contractor will meet the company's HSE requirements and will manage the identified HSE risks while performing the applicable activities in the project execution stage that generally will include detailed engineering and design, construction and commissioning. This includes the method statements for various critical activities to address all pertinent HSE aspects during project job execution.

7.3 Hazard management plan

Hazard Management comprises of three following parts:

- Identification/recognition;
- Assessment and analysis;
- CAPA (corrective and preventive action).

Through walkthrough and site visit, inspections conducted and identified non-compliance and sub-standard practices. Situations or conditions leading to serious consequences and prone to injuries are brought into considerations. Communicating inherent potential hazards in various project critical activities to the concerned worksite supervisors/workers involved in the job, is one of the effective ways of managing hazards and thus preventing incidents at worksite. HSE-NK team takes special consideration on the above and ensured about the proper hazard identification, control through assessment and communication to the concerned workers through toolbox talk, Supervisor briefing. Surprised checks are being made randomly at some of the worksites having hazardous activities to ensure compliance. Suitable hazard briefings provided to available workforce wherever it was needed/necessary.

7.4 HSE hazard communication

Internal HSE hazard communication and management processes are focused on maintaining open lines of communication within and between project operational and support groups. Regularly scheduled meetings at all levels of the organization allow two-way dialogue to improve communication on HSE matters. In addition to regular personal communication between KOC (controlling teams) and the contractors' workforce, KOC maintains a number of communication tools and reference sources to provide employees with on-demand HSE information.

7.5 Exclusive HSE awareness

Annual program on "Onsite HSE awareness" sessions exclusively for project contractors' personnel at their worksites on monthly basis covering relevant/applicable HSE topics (namely, working at height, excavation, mobile equipment, work permit/JSA, confined space, electrical safety, driving safety, etc.) covering all high priority/project critical activities apart from theme-based campaigns. Last year more than 4,000 contractors' workers were covered under onsite HSE awareness programme at various levels. This has ensured more clarity on the procedures and effective compliance.

7.6 Sharing HSE best practices/lessons learned

Communicating HSE best practices, lessons learned, safety flashes/messages, HSE circulars, etc. regularly amongst all project contractors through various forums including publishing theme based HSE Flash in company's portal, screen saver, HSE meetings, etc. with an objective of enhancing HSE awareness as continual process.

7.7 Special HSE monitoring

Physical examinations and surveys at worksites are being conducted randomly especially for the critical activities, namely, confined space entries, heavy lifting/shifting activities, etc. In the above process, field compliance audits are also carried out considering the potential hazard identification and suitable preventive/protective measures taken including the proper permit issuance as per the applicable procedures. Immediate corrective measures being taken wherever deviations are noticed.

7.8 Theme-based audits

Conducting special audits on job-specific activities (namely, permit to work, excavation, etc.) to evaluate the compliance level in order to improve the contractors' and sub-contractors' HSE performance [2].

8 RESULTS GAINED AND ADVANTAGES

Addressing HSE issues with a proactive risk-based approach during project execution activities, it not only ensures compliance as per KOC requirements, but also offers significant opportunities to improve HSE performance. There has been significant improvement in the HSE performance of North Kuwait Directorate by following the above HSE best practices in managing project risks at various worksites. Year wise comparison of HSE lagging indicators of NK areas is graphically given in Fig. 5.

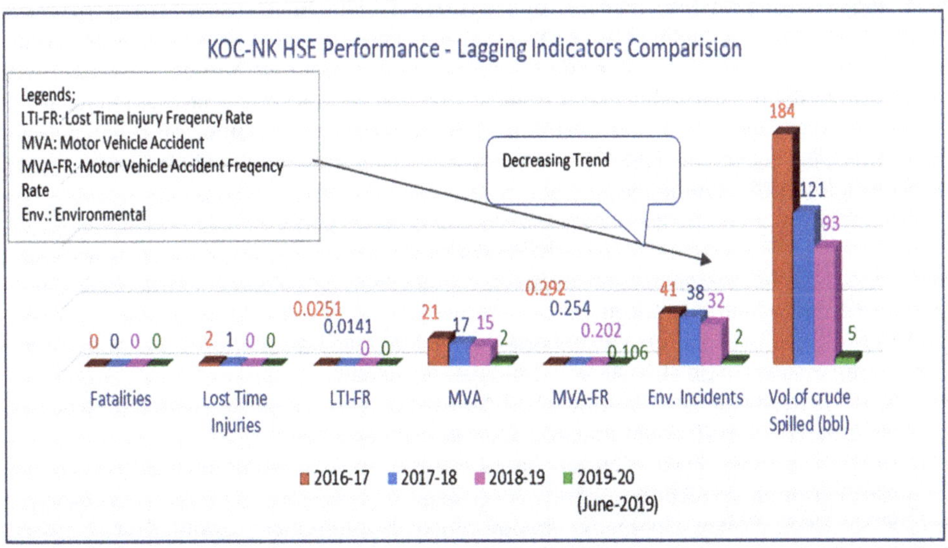

Figure 5: HSE lagging indicators comparison.

Some of the benefits are as follows:

- Zero lost workday cases (of contractors) in NK Fields since 19th April 2017.
- Provided a mechanism for risk-based approach on contractors HSE compliance to the controlling teams and contractors management.
- Performance of NK contractors is benchmarked, which is utilized to enhance their Performance compliances.
- Benchmarking encourages healthy competition amongst NK contractors to strive for better HSE performance.
- Positive field HSE compliances has increased substantially, which resulted in positive HSE statistics for NK fields.
- Provides a platform for the controlling team to effectively monitor contractor HSE performance compliance level.
- Effective tool for competent workforce hiring, promotions, work assignments, rewards/penalties of contractors.
- Provides a structured approach focusing on contractors' strategic performance objectives.

9 CONCLUSION

"Risk management practices" followed during project execution activities is one of the very important approaches for ensuring incident free activities and improved HSE performance. The compliance gaps and positive observations identified through this process are used as a basis for the analysis and subsequent recommendations, in order to improve upon HSEMS procedural compliance. It is proved to be an effective tool for monitoring and management of critical high risks posed by various activities using systematic and structured approach. However, maintaining the consistency in the risk-based approach and supporting its activities is not an easy task. Hence, continuous involvement and deliberations of the team-members with appropriate logistic supports rendered by the HSE-NK leadership is very much essential and vital in order to manage the critical risks in the desired manner.

With the relentless efforts rendered by the HSE-NK team, there are visible improvements in HSE compliance including hassles-free and incident-free activities during entire project job execution period at various project sites at North Kuwait Directorate. The long-term intent is to build project findings to document best practices, lessons learned, and examples of various documents to benefit KOC and the contractors in terms of benchmarking, evaluation and enhancing HSEMS performances. This kind of initiative clearly signify organization's HSE commitment.

REFERENCES

[1] KOC, HSE Management System Guide, A framework for company's HSE Management System, KOC internal document.
[2] KOC, HSEMS Performance Reports of Contractors, HSE performance expectations during job execution at KOC, KOC internal document.
[3] KOC, HSE Management System (HSEMS) Procedures, A management tool to achieve the Company's HSE commitment, KOC internal document.
[4] KOC, "MyHSSE" Data, A system based software application for managing all relevant HSE performance records.
[5] KOC, HSSE Policy, Company's philosophy and commitment towards Health, Safety, Security & Environment, KOC internal document.

PRACTICALITY OF RISK ANALYSES OF ESTABLISHMENTS

HELMUT KÖRBER
APC Angewandte Physik Consulting AG, Switzerland

ABSTRACT

Quantitative risk analyses require high expenditures from a skilled workforce (many person-years), which is mostly unaffordable for an SME industrial establishment. Therefore, a much less workforce-consuming methodology – SQP (Semi-Quantitative Procedure) – was developed. SQP consists of two parts, TRA (Tabular Risk Analysis) and ACCS (Analysis by Conservatively Covering Selection). The risk R is the product of the severity of a damage S of an accident and the probability P of its occurrence. Instead of quantitative numerical values for S and P, SQP is using risk parameters Sn and Wn, similar to VDE/VDI 2180 or SIL. From this, a risk indicator Rn is determined. The method provides a set of differentiated levels of risk. Risk analyses using SQP have been carried out for several industrial establishments being subject to Seveso III Directive upper-tier group. In all cases, it was possible to uncover those (in general very few) installations of the site of operation generating a risk reasonably above the average. The introduction of SQP in the companies has been easy due to the similarity and the experience of the engineers with SIL. Instead of a workload of person-year(s) for a risk analysis of an establishment, only some two person-weeks are required.

Keywords: risk assessment, industrial establishment, upper-tier, probabilistic analysis, Seveso III, semi-quantitative risk analysis, SIL, severe accident, low workload, practicality.

1 INTRODUCTION

Following existing laws, directives, regulations, standards and rules, industrial establishments implement well-defined organizational measures and sophisticated technical systems to guarantee the safe operation of in their plants. The deterministic way of thinking results in the conclusion that, if all is carried out properly and correctly, accidents are and will be prevented. However, experience shows that accidents still occur, nevertheless. This can be evaluated with probabilistic methods, which eventually leads to risk analyses.

The evaluation of complex technical systems and systems under the perspective of risk is not really a new topic. Already since the 1970s, a growing part of the technical and scientific society has become aware of quantitative risk analyses, especially those carried out for nuclear power plants [1], [2]. The labour effort required was huge. In all investigations, it did sum up to many person years of workload of skilled scientists and engineers.

Therefore, quantitative risk analyses were initially carried out for conventional technical installations and plants in exceptional cases only. The reason for risk studies was always a very special boundary condition, for instance if a hazard potential was regarded from the plant operator in contrary to the supervisory authority, or in case of highly controversial projects such as the erection of a hazardous waste incineration plant in a metropolitan area. Another example is the expansion project (additional runway) of Frankfurt/Main airport that led to an extra hazard for a chemical installation as each aircraft in descent and final approach had to cross the site at low altitude [3].

It is understandable that such a big deployment of qualified labour appears to be disproportionately high – at least for small and medium enterprises (SME) operating an installation subject to upper-tier requirements of Seveso III Directive [4]. It is also questioned whether these additional expenditures for skilled staff required to carry out a quantitative risk analysis is not beyond the limit of the reasonableness for a company being exposed to international competition.

WIT Transactions on The Built Environment, Vol 189, © 2019 WIT Press
www.witpress.com, ISSN 1743-3509 (on-line)
doi:10.2495/SAFE190131

2 SKEPTICISM AND BENEFITS

Parts of the population became aware of the risk studies. In particular, in German-speaking countries, it had to be observed that both the calculation with probabilities and the calculatory consideration of dead and injured resulted in mistrust, suspicion and also in full rejection [5]. Even engineers in companies or supervisory authorities who had to deal with the safety of industrial plants did show a reduced confidence. Some are still sceptical today.

2.1 Risk analysis – reasons for scepticism

A one-dimensional explanation for the scepticism cannot be given. In any case, the accumulated experience with regulations, directives and laws plays a key role. This experience has been gained within a period of several generations and covers not only the technical sphere but also the social side of life, after which everything works and nothing goes wrong, if you only follow exactly the given rules. This is almost always true. However, experience shows, too, that despite the best precautions and meticulous compliance with all regulations, incidents can still not be ruled out with absolute certainty.

For most people it is (subjectively) a difference, whether they expose themselves to a risk or whether they are exposed to it from outside, from third party. High risks, for example due to dangerous sports, unhealthy diet, smoking, alcohol and drug consumption are much more likely to be accepted than the possibly low risk that emanates from a neighbouring chemical factory. The risk of participating in road traffic is mentally rated as low, if we drive ourselves. On the other hand, the risk is considered as high, if we are threatened by traffic (from others).

Moreover, it can be observed that the term risk is sometimes not used correctly – in particular in conjunction with negative events such as industrial incidents and accidents. That is why many plant operators avoid the use of the word "risk", because it can be misused from opponents of technical installations. Even if a plant is particularly safe and emanates an extremely low (residual) risk only, the term risk is often unconsciously or consciously, but mistakenly – used as a synonym for "danger", "endangerment", "accident", "hazard" etc. Like this, an impression arises that the installation is very dangerous.

Furthermore, the thinking in terms of probabilities or frequencies of events to be expected is alien to many people, mainly if it is a question of low probabilities. This can be explained by our horizon of experience which covers about 80 to 100 years (three generations). We still can imagine that a major fire occurs in an average of once every 10 years, or that, say, a residential building is destroyed by a gas explosion in the own city approximately every 25 years.

On the other hand, we cannot differentiate with regard to low probabilities. Events being expected at an average frequency of once every 500 years are of the same category as those occurring once in 50,000 years.

2.2 Reasons for conducting risk analyses

There are several reasons for carrying out risk analyses. The most important are:

- Risks are comparable; even if they arise from different potential threats. Thus, the risk to which a resident is exposed from a facility nearby can be compared with natural risks such as earthquakes or floods and also with the risk the resident is responsible for by himself and he decides to take it voluntarily. Unless preconceived opinions oppose it, the comparison of risks can contribute to the acceptance of technical systems.

- By the help of a systematic risk assessment of an entire industrial installation, those (in general few) sub-areas/components can be identified causing an increased risk. This makes it possible to purposefully implement measures of risk-reduction and thus reducing the remaining total (residual) risk the whole establishment.
- Installations being assigned to upper-tier establishments corresponding to Seveso III Directive have to carry out risk evaluations.

Seveso III [4] is a Directive of the European Union (EU) for the control of major-accident hazards with dangerous substances. It is in force in all member states in form of a law or legitimated regulation that represents a transposition of the Directive into the legal framework of each EU state. Seveso III is also relevant for the European Economic Area (EEA).

Corresponding to Seveso III Directive, a determination and analysis of the risks of accidents is required as well as the means to prevent them, with a detailed description of the scenarios of possible major accidents and their probability under which they occur. Additionally, an assessment of the extent and severity of the consequences of major accidents identified is necessary, and all this information has to be documented in the safety report (Seveso III Directive [4], Annex II, 4).

Furthermore, a safety management system (SMS) is required and the organisation of the establishment with a view to the prevention of major accidents. The SMS must take into account the aspects of identifying and evaluating major hazards through adoption and implementation of procedures for systematically identifying major hazards arising from normal and abnormal operation including subcontracted activities and the assessment of the likelihood and severity of such accidents (Seveso III Directive [4], Annex III, b, ii).

As already mentioned, a quantitative risk analysis is very person-power consuming. Therefore, the methodology SQP (Semi-Quantitative Procedure) was developed with the objective of being able to carry out a simplified and nevertheless significant and meaningful systematic risk assessment of industrial establishments. SQP consists of two parts, the TRA (Tabular Risk Analysis) and ACCS (Analysis by Conservatively Covering Selection). First, the two components of SQP will be explained below, and the procedure will be presented. This is followed by an experience report on the introduction of SQP in practice.

3 THE SEMI-QUANTITATIVE PROCEDURE (SQP)

3.1 Tabular risk analysis (TRA)

A systematic safety evaluation must cover the entire industrial establishment. The approach of SQP is to subdivide the establishment into associated section or sub-systems respectively to carry out the TRA. Each sub-system is evaluated with regard to existing potential hazards and the existing preventive measures against possible malfunctions.

In general and mostly, an establishment is composed of several sections which can be defined as subsystems. An example, a chemical plant is shown (Fig. 1). The site has an unloading area (Section 1) where the delivered raw materials are taken and transferred to the entrance storage (Section 2). The manufacturing of products takes place in two steps, in process lines for producing the raw substance (Section 3) and the cleaning system to purify them (Section 4). Finally, the end products are stored in a warehouse or outlet tanks for being shipped to the customer (Section 5). These five sections can be used for TRA.

In the case of a crude oil refinery or a petro-chemical installation like an ethylene plant, there is another option to determine the sub-systems. Most of these facilities are equipped with system to section-wise separate parts of the installation from the rest. This is realized

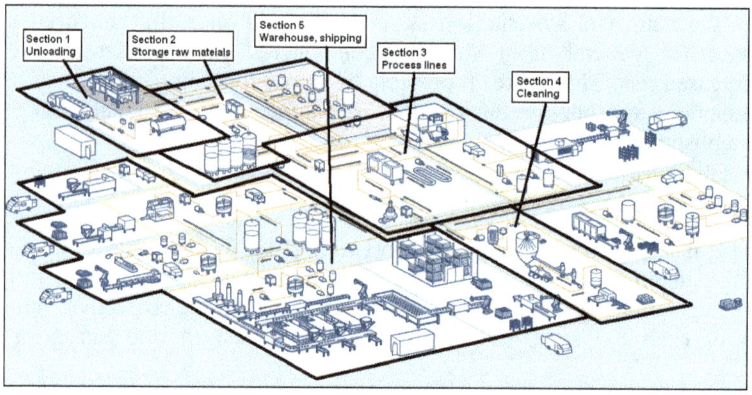

Figure 1: Site of an industrial establishment.

through sealing-off the section in question, containing/locking-in the materials (containment) and taking care for a controlled emergency release or discharge. If so, the sections of such an Emergency Seal-off and Release System (ESRS) can be defined for the tabular risk analysis as sub-systems.

In each sub-system there are numerous technical and organisational measures to ensure a safe operation. Nevertheless, it cannot be ruled out that failures of components or faulty actions of persons occur resulting in a malfunction or in a major accident. It is possible that a vulnerability of an instrument or the weak spot of an apparatus or any other deviation from lay-out and regular operation triggers an accident. This is to find out.

For this purpose, all imaginable chains of events or scenarios must be considered and analysed. A systematic compilation is inevitable. All conceivable root causes must be captured that reasonably can lead an accident. Each event sequence uncovered is listed up in a table (Table 1). Based on this, all these sequences are analysed and evaluated under the aspect of risk. (This investigation is carried out for all sub-systems.) The procedure of carrying out a TRA is given below; Table 1 is to be used as guideline.

Table 1: Working table for carrying out a tabular risk analysis (TRA).

Sub-system	Case and component selected	Potential hazard	Conceivable root cause/ triggering event	Conceivable event sequence/ potential extent of damage	Existing measures to prevent/limit the damage	Risk parameter		Risk indicator	Remark
						S	P	R	

In the sub-system being subject to an analysis (column 1), all components or cases are to be considered (column 2) that might be the reason for triggering an accident. The potential hazards or pessimistic consequences of these disturbances (column 3) and the triggering events (column 4) are compiled and listed. Based on this, each scenario leading to the accident is explained in keywords representing the conceivable chain of events and the potential extent of the damage (column 5). Columns 2 to 5 briefly illustrate the assumed accident and the possible extent of damage under the assumption that no or only insufficient provision (both technical and organizational) was made against the accident.

For each event sequence the existing measures are listed that shall prevent an accident and, should it occur nevertheless, limit its effects (column 6). These can be technical devices/components but also organizational measures. Examples: the automated addition of an inhibitor into a reactor to stop an excursion, to activate ESRS in case of a problem in the concerned section, an alarm system to indicate the release of hazardous gases or to start emergency response plan, an automated initiation of installed extinguishing devices to fight a fire. These existing measures shall (and, in general, do) take care of an effective prevention that a malfunction or another triggering event cannot initiate an accident.

If, however, the analysis uncovers safety-related weak points, it is quite likely that this results in an accident. Depending on the severity of damage to be expected and the probability of the occurrence, the risk for this accident can then be calculated or assessed (see next section). If necessary, measures have to be taken to reduce the risk.

3.2 Calculating the risk

To assess and the risk of accidents, the severity of the damage is required as well as the likelihood that it will occur. The risk R is defined as the product of the severity of damage S and the probability of the occurrence W (see [1], [2], and Seveso III [4], Article 3, 15):

$$R = S \cdot W.$$

In a quantitative risk analysis, the severity of damage S has to be specified. An approach is to use the expected property loss for S (e.g. in €), or the personal harm (number of injured or deaths). S can also be expressed in an indirect harm like the concentration of a toxic gas at the point of impact (e.g. in mg/m^3 breathing air). Environmental damages can be quantified, too (e.g. in g/m^2 contaminated agricultural land) as well as image damages (e.g. number of negative reports). The user can also define the severity of damage by himself.

The probability is regarded as the average expected frequency of the event per year. For example, if an assumed accident A is expected on average twice a year, the probability for this is $W_A = 2\ a^{-1}$. On the other hand, if accident B is expected to occur once every 10 years on average, this means 1/10 times a year; thus $P_B = 0{,}1\ a^{-1}$.

If statistics show that, say, out of 200 chemical plants sited in the lane of approach of airports, one plant is hit by a plane crash on average every 50 years, then the probability W_C for such an event is for each of these chemical plants $W_C = 10^{-4}\ a^{-1}$. Hence, an impact of an aircraft crashing down on the site can be expected per plant one time every 10,000 years.

In case, this crash will cause a property damage S_C of €80 million (and assumed that no personal or environmental damages are to be expected), the risk R_C can be easily estimated. It amounts to 80 million € times $10^{-4}\ a^{-1}$, i.e. €8,000/a. If the plant operator wants to take out insurance against the consequences of such a crash, the insurance company would set a premium on this basis, i.e. not charge less than €8,000 annually.

3.3 Approximation method SQP for risk assessments

As performing a quantitative risk calculation is possible but very costly, an approximation method has been developed. It results from practical working experience in industrial upper-tier establishments and is safety analysis in Semi-Quantitative Procedure (SQP). SQP does not use accurate quantitative numerical values for severity of damage S and the probability of occurrence W, because each term requires an extended portion of time and is not very precise, nevertheless. Instead of this, SQP is based on risk parameters Sn for the extent of damage and Wn for the probability. From these, the risk is determined as risk parameter Rn indicating the level of risk.

The parameter Sn for the severity of damage is defined in very close accordance with the widely used standards of VDI/VDE 2180 [6] or SIL [7] considering a graded level of damages S1 to S4 (Fig. 2). However, the severity of damage S0 was added in order to be able to extend the differentiation of damage grades. S0 symbolizes lower values for property damages and reduced or not relevant injuries of persons.

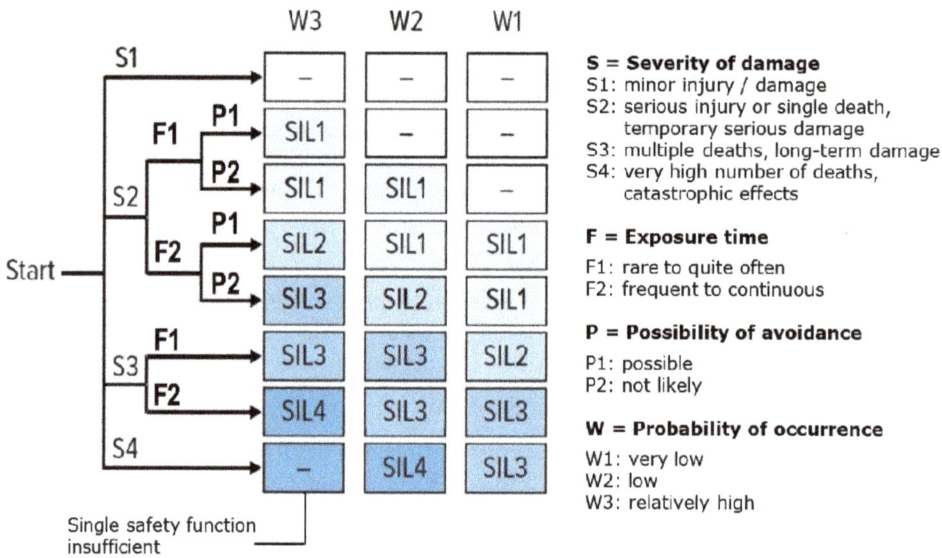

Figure 2: Risk graph corresponding to IEC 61508.

A severe accident in industrial installations can not only result in injuries of persons and lost values of property. It can also have an impact to the ambient. An example for this is the accident in the chemical plant of the company Industrie Chimiche Meda SA (ICMESA) in northern Italy causing an environmental disaster. The accident happened in 1976. Highly hazardous dioxin was spread over into the ambient. Agricultural land was devastated, and several neighbouring towns and villages were hit. Seveso was the community most affected. As another and additional outcome, the image of the company was ruined.

An accident in general casts a bad light on the related operation. Therefore, the extent of the damage does not only include personal injury, environmental damage and property damage, but also expected negative effects on the image. Hence, graded levels for environmental damages are included in SQP as well as expected negative effects on the image of the plant operator. Table 2 summarizes the definitions of the risk parameters Sn and provides the key words for the analysis.

Table 2 also indicates how the parameter Sn is to be applied in investigations according to Seveso III Directive. The extent of damage is relevant for the evaluation of accidents, defined in the Directive as "serious danger to human health or the environment" (Seveso III Directive [4], Article 3, 13). If this criterion is not met, it is not a major accident corresponding to the definition of Seveso III, but an incident. Linking this with the risk parameters Sn for the extent of damage according to Table 2, the result is as follows:

S0 = no relevant injury to persons, negligible environmental damage
= no accident,
S1 = minor injury, one person, minor harmful environmental damage
= no accident (as no "serious danger" corresponding to Seveso III),
S2 = serious injuries or single dead, short-term major environmental damage
= accident (as criterion "serious danger" corresponding to Seveso III),
S3 = seriously injured, several dead, long lasting major environmental damage
= major accident,
S4 = high number of deaths, irreversible environmental damage
= very large accident (catastrophe).

Table 2: Definition of risk parameter Sn for severity of damage.

Risk parameter for severity of damage	Extent of damage in case of effects on:			
	Persons	Environment	Image	Property values*
S0	No relevant injury	Negligible small	No effects	< 25 k€
S1	Minor injury, one person	Minor harmful effects	Negligible effects	< 250 k€
S2	Serious injuries, several persons, irreversible injury or single death	Short-term major effect	Short-term negative effects	< 2.5 Mio €
S3	Serious/irreversible injuries, many persons, multiple deaths	Long lasting major damage	Major negative impacts	< 25 Mio €
S4	Very high number of deaths	Long lasting big/irreversible damage	Long lasting negative impacts	≥ 25 Mio €

*k€ = €1,000, Mio € = €1,000,000.

The methodology SQP also uses graded parameters Wn for the probability of occurrence of the undesired event. Consequently, Wn is defined in close accordance with VDI/VDE 2180 [6] and DIN EN 61508 and IEC 61508 [7] (see Fig. 2). However, the grading of W1 to W3 is too crude for the practical application in risk assessment of industrial establishments. In particular, there is no option to also evaluate extremely rare events (that might possibly cause a huge damage). Therefore, the parameter W0 was added. This results in a span of W0 (extremely low) to W3 (relatively high) for the probability parameter Wn. Table 3 summarizes these parameters with a brief explanation of their meaning and typical numerical values for the respective probability.

These probability parameters correspond with quantitative probabilities: If an accident (e.g. large fire) is to be expected at least once in the course of service life of an installation, Table 3 indicates the use of W2 (= low probability of occurrence). Assuming a lifetime of the plant of 20 years (the technical service life is to be considered and not the economic life corresponding to the depreciation period), the probability W_{Fi} for a fire in this plant is $W_{Fi} = 1/20\ a^{-1}$. Hence, the parameter W2 corresponds to a quantitative probability of

$$W_{Fi} = 5 \cdot 10^{-2}\ a^{-1}.$$

Table 3: Definition of risk parameter Wn for the probability of occurrence.

Risk parameter for probability of occurrence	Probability of occurrence		
	In short formulation	As imaginable frequency	As typical order of magnitude
W0	Extremely low	The incident is not expected in the course of service life of the installation, but is elsewhere in the range of the possible	$2 \cdot 10^{-3}\ a^{-1}$
W1	Very low	The incident is to be expected at the most once in the course of service life of the installation, elsewhere it has already taken place	$1 \cdot 10^{-2}\ a^{-1}$
W2	Low	The incident has to be expected at least once in the course of service life of the installation or has already taken place	$5 \cdot 10^{-2}\ a^{-1}$
W3	Relatively high	The event has to be expected several times in the course of service life of the installation, respectively has already taken place	$2 \cdot 10^{-1}\ a^{-1}$

3.4 Determination of the risk indicator

Using the parameters S0 to S4 for the severity of damage and W0 to W3 for the probability of occurrence of the undesired event, a parameter Rn for the risk is determined. For this, a formal production of Sn and Wn is carried out – in close alignment with the quantitative risk calculation (risk = severity of damage multiplied by the probability):

$$Rn = \{ Sn \} \cdot \{ Wn \}.$$

Like this, the risk can be easily "calculated" and expressed in different levels. This is used in the tabular risk analysis (Table 1). The parameters Sn and Wn are determined for each of the conceivable event sequences well as the risk indicators Rn. All is entered into columns 7 to 9 of the table. As numerous event sequences must be considered in the TRA of an industrial establishment, many risk indicators Rn appear in Table 1, correspondingly. In an evaluation of conceivable accidents these levels of risk are used:

- R0 – R1: extremely low to low risk,
- R2 – R3: medium risk,
- R4 – R6: big risk that should be reduced in medium to short term,
- R7 – R12: very high risk that cannot be tolerated.

This can also be displayed in a risk matrix (Fig. 3).

4 ANALYSIS BY CONSERVATIVELY COVERING SELECTION (ACCS)

Carrying out a TRA (tabular risk analysis) of an industrial establishment with all its sub-systems and components, results in the examination of many conceivable event sequences that might result in a major accident. Even in medium-sized installations of SME, the number of scenarios to be considered can sum up to several hundred. It is possible to reduce this number considerably by using the ACCS method (Analysis by Conservatively Covering Selection). Nevertheless, the safety of the establishment can be systematically checked.

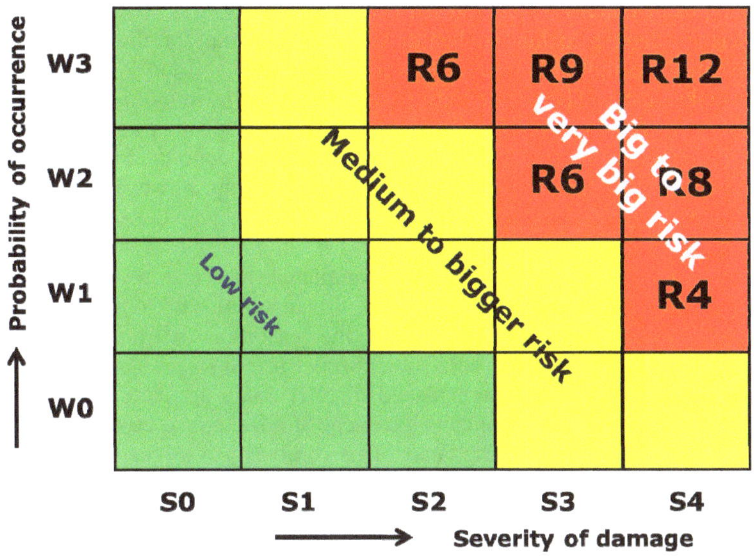

Figure 3: Presentation of the results in a risk matrix.

4.1 The ACCS method

ACCS was developed on the basis of the Failure Mode and Effects Analysis (FMEA) [8], which had been derived from the Failure Effect Analysis (DIN EN 25 448). As ACCS is a systematic safety assessment, it complies with the requirements of Seveso III. Following the ACCS method, those accidents with the most serious negative consequences can be distilled from the entire big spectrum of conceivable accidents. The procedure is as follows.

Starting point is the group of sub-systems of the establishment. In each of these systems, the units and components with the highest potential of accident damages are selected conservatively and listed in Table 1, columns 2 and 3. This is, for instance, the vessel with highest pressure, or the reactor with the more dangerous substances, the tank with the biggest quantity of hazardous materials, the pump with a higher head, the filter with the most toxic dust, the liquid with biggest potential for health hazards, etc. Likewise, from the variety of conceivable triggering events the most conservatives are selected.

All these selections are conservative. Therefore, the sequence resulting from a specified malfunction or a triggering event and the corresponding chain of events (columns 4 to 6) also conservatively cover other comparable cases.

Based on this, the risk indicator is derived for each conceivable accident. At the end of this investigation for the entire industrial establishment covering all sub-systems, a list of the relevant individual risks is displayed that exist in a plant. By comparing the risk indicators and analysing their causes, it is possible to specify and stipulate those components, sub-sections, actions of persons, etc., generating the biggest overall risk.

4.2 Example of a risk analysis

The application of SQP as a TRA together with ACCS is shown below using the example of a chemical plant – at first by considering the sub-system 2 "in-tank farm". Here, there are various types of materials to be stored in tanks. Ammonia is the substance with by far the most serious danger to human health and the potential for the most serious major accident. Within the tank farm, there several components filled with ammonia (pipes, containers, tanks, etc.). The conservative selection results in taking the NH_3 tanks as components for the further investigation, due to the biggest quantity ammonia.

There are four NH_3 tanks, all of the same size and the same thermo-physical data with regard to operation. All tanks are placed in a collecting tray; however, they differ in age: Two of the tanks went into operation seven years ago, whereas the other two are over 30 years old. Therefore, the latter are conservatively selected for the TRA (see Table 4). These two older ammonia tanks are largely but not completely identical in construction. Both are made of 1.0425 steel (other name: H-II). Today, this material is no longer recommended for the storage of pressure-liquefied NH_3. One of the reasons is that there are more suitable steels which, moreover, have better cold-tough properties, too.

The two old tanks (A and B) are placed directly aside an access road to one of the factory buildings. Therefore, the conceivable event of a fire has to be considered that might arise in the immediate vicinity of the tanks. This can result in a bursting of one or both of the tanks. If so, the massive release of NH_3 conservatively covers other, minor releases (such as leakage through break of a supply pipe, leaking flange, and the like). The risk potential in the wake of a fire and, as well, a brittle fracture affects both old tanks about the same. On the other hand, the risk of stress corrosion cracking is in issue mainly to the older of the two old NH_3 tank because it had not been stress relief annealed.

In summary, three conceivable accidents have to be considered, a “fire in front of the tank”, “stress corrosion cracking” and “brittle fracture at low temperature”. In any case, the extent of damage corresponds to a major accident with several to many deaths be expected due to the massive release of NH_3. The risk parameters Sn were derived as S3 to S4. The existing measures and provisions for preventing such great damage lead to an evaluation of the risk parameters Wn for the probability of W1 to W2 (Table 4).

This results in the risk indicators of R8 (in the case of a fire next to the tank), R5 (stress corrosion cracking) and R7 (brittle fracture at low temperature), see Table 4.

The risk assessment had not only to be carried out for the sub-system 2 “tank farm” but also for all others, i.e. for the entire establishment. This was also preformed using SQP method with TRA and ACCS. The investigation resulted in a set of risk parameters Rn for some 35 conceivable event sequences and accidents, all of them conservatively covering other event sequences and respective accidents with lower extent of damage. The outcome was that – with the exception of the tank farm – only quite low risks were determined with characteristic parameters of R1 and R2, in some few cases also R3 was found, but nowhere R4 or higher. Thus, the question could be answered as to whether the operation of the plant is connected with a high/very high risk and, if so, which section it emanates from: Yes, there is a considerable risk, and this results from the operation of the NH_3 tanks.

5 EXPERIENCE WITH RISK ASSESSMENTS ACCORDING TO SQP

The introduction of the SQP method for risk analyses in companies is considerably facilitated by the fact that the utilisation of S and W is strongly based on the regulations of VED/VDI 2180 [6] and SIL [7]. All relevant engineers are familiar with this method. That is why they easily can apply the SQP risk approximation method. But there is also a disadvantage: The determination of the risk indicator Rn can also result in R0 – in case of superficial speech, apparently a zero risk. However, this is wrong; this impression must not occur.

There is no zero risk. A risk can be extremely low – but never zero.

Table 4: Tabular risk analysis according to SQP with ACCS method for the safety evaluation of the sub-system tank farm of a chemical plant.

Sub-system	Case and component selected	Potential hazard	Conceivable root cause/ triggering event	Conceivable event sequence/potential extent of damage	Existing measures to prevent/limit the damage	Risk parameter*		Risk indicator	Remark
						S	P	R	
2 NH_3 tank farm	2.1 Tank A, built in 1981	Bursting of tank, extensive release of NH_3	Fire in front of tank and collecting tray	Tanker catches fire in front of NH_3 tank Heat radiation Bursting of tank Release of NH_3 Dispersion to residential area Many dead	Fire alarm through staff Site fire brigade Alarm and emergency response plan (AERP)	S4	W2	R8	Fire alarm not automated
2 NH_3-tank farm	2.2 Tank A, built in 1981	Bursting of tank, release of NH_3	Stress corrosion cracking	Tank not stress relief annealed Crack formation Leakage Release of NH_3 Several dead High image-damage	Site fire brigade Alarm and emergency response plan (AERP)	S3	W1 to W2	R5	
2 NH_3-tank farm	2.3 Tank B, built in 1987	Bursting of tank, release of NH_3	Brittle fracture at low temperature	Tank material limited cold ductility only No pressure reduction at low temperature Brittle fracture Release and disperse of NH_3 Several, or may be many dead High image-damage	Site fire brigade Alarm and emergency response plan (AERP)	S3 to S4	W2	R7	

*Severity of damage S and probability of occurrence W:
S0 = minor property damage, no injured persons, S1 = moderate damage, S2 = accident, S3 = major accident, S4 = catastrophe,
W0 = extremely low, W1 = very low, W2 = low, W3 = relatively high.

Table 5: Tabular risk analysis of the sub-system tank farm according to SQP method after implementation of measures to improve safety.

Sub-system	Case and component selected	Potential hazard	Conceivable root cause/triggering event	Conceivable event sequence/potential extent of damage	Existing measures to prevent/limit the damage	Risk parameter*		Risk indicator	Remark
						S	W	R	
2 NH_3 tank farm	2.1 Tank A, built in 1981	Bursting of tank, extensive release of NH_3	Fire in front of tank and collecting tray	Tanker catches fire in front of NH_3 tank Heat radiation Bursting of tank Release of NH_3 Dispersion to residential area Many dead	Road closed Automated fire alarm Tank sprinkling Foam extinguishing system Site fire brigade AERP	S4	W0 to W1	R2	
2 NH_3-tank farm	2.2 Tank A, built in 1981	Bursting of tank, release of NH_3	Stress corrosion cracking	Tank not stress relief annealed Crack formation Leakage Release of NH_3 Several dead High image-damage	Tank is stress relief annealed Annual material inspection Q/A iof NH_3 purity Site fire brigade AERP	S3	W0 to W1	R2	
2 NH_3-tank farm	2.3 Tank B, built in 1987	Bursting of tank, release of NH_3	Brittle fracture at low temperature	Tank material limited cold ductility only No pressure reduction at low temperature Brittle fracture Release and diserse of NH_3 Several, or may be many dead High image-damage	Automized pressure reductuction Operating instruction for pressure/temperature monitoring of tank Periodical checks for possible cracks Site fire brigade AERP	S3	W1	R3	

*Severity of damage S and probability of occurrence W:
S0 = minor property damage, no injured persons, S1 = moderate damage, S2 = accident, S3 = major accident, S4 = catastrophe,
W0 = extremely low, W1 = very low, W2 = low, W3 = relatively high.

There are also other risk assessment methods with parameters considering graded levels for the degree of damage and probability of an event – ARAMIS [9] is an example. The main difference is that these methods do not use a similarity to SIL. Therefore, an introduction and efficient application is not as fast as the SQP methodology with TRA and ACCS, and generating results is more time consuming. (That's exactly why SQP was designed similar to SIL.) A quantitative risk analysis is to be compared with SQP under workload aspects: Instead of consuming person year(s) of skilled workforce, SQP enables a medium-sized installation to be risk analysed with limited workload of two to three person weeks only.

Once the spots causing high risk are known, this can be reduced, for instance by retrofitting. Possible variants/alternative solutions for the improvement can already be evaluated in the phase of project planning through SQP method and TRA (tabular risk analysis). This is shown using the example of the chemical plant with NH_3 tank storage (Table 4): Improved measures for fire detection and fire-fighting, the stress-relieving annealing of tank A, various adjustments in metering an control, better monitoring and inspections measures significantly reduced the probability of occurrence of accidents (see Table 5). The risks thus identified are now significantly lower and tolerable.

Risk analyses are powerful tools to prevent severe accidents in industrial installations. There are various approaches to carry out these investigations. Although the basic risk equation is the same, the risk assessment applications are different. A harmonized method, defined as a standard, would help to apply risk analyses more comprehensively and, thus, reducing losses and damages. This is stated as a final recommendation.

REFERENCES

[1] Rasmussen, N.C. et al., Reactor safety study. An assessment of accident risks in US commercial nuclear power plants. *WASH-1400,* NUREG-75/014, 1975.

[2] Bundesministerium für Forschung und Technologie (Herausgeber), Deutsche Risikostudie Kernkraftwerke, Phase B. Verlag TÜV Rheinland, Köln, 1990. ISBN 3-88585-809-6. (In German.)

[3] Jochum, Chr., Quantitative Risikobewertung für Chemieanlagen bei Land Use Planning – Fallstudie Störfall durch Flugzeugabsturz. Meinungs-und Erfahrungsaustausch für Sachverständige nach §29a BImSchG in Dessau. *Sammlung der Präsentationen,* Warm Engineering (Herausgeber), Freilassing, 23. März 2006. (In German.)

[4] Directive 2012/18/EU of the European Parlament and the Council of 4 July 2012 on the control of major accident hazards involving dangerous substances, amending and subsequently repealing Council Directive 96/82/EG, (Text with EEA relevance). *Official Journal of the EU* (2012) L 197, pp 1–37.

[5] Rupohl, G., Das misstrauen in der technikdebatte (Mistrust in the technology debate), *KIT Scientific Publishing,* ISBN 9782821894884, 2010. (In German.)

[6] VDI/VDE 2180, Sicherung von Anlagen der Verfahrenstechnik mit Mitteln der Prozessleittechnik (P–LT). Beuth Verlag: Berlin, 2007. (In German.)

[7] DIN EN 61508 & IEC 61508, *Functional Safety of Electrical/Electronic/ Programmable Electronic Safety-Related Systems*, Beuth Verlag: Berlin, 2011.

[8] DIN EN 60812, *Failure Mode and Effects Analysis (FMEA)*, Beuth Verlag: Berlin, 2015.

[9] Debray, B. & Salvi, O., ARAMIS Project: An integrated risk assessment methodology that answers the needs of various stakeholders. *WIT Transactions on The Built Environment*, vol. 82, WIT Press: Southampton and Boston, pp. 265, 2005.

Author index

Maritime Transport

Edited by: ***G. PASSERINI****, Marche Polytechnic University, Italy and* ***S. RICCI****, University of Rome 'La Sapienza', Italy*

Maritime transport faces multiple challenges, therefore it requires an interdisciplinary approach in order to respond efficiently to the interaction between diverse agents. This book presents interdisciplinary research, as well as operational experiences, which contribute towards the development of the field.

The substantial growth of maritime shipping has resulted in large quantities of good and products being transported around the world, creating a demand for innovative solutions for infrastructure and fleets. A further increase in the scale and the size of container carriers and passenger cruisers also requires special facilities. These developments have presented a challenge for different types of technologies, as well as operational systems.

Maritime transportation ought to be integrated with other services such as railways, roadways and airports and, in some cases, river and canal traffic in order to achieve maximum efficiency. It also needs to respond rapidly to the economic and political circumstances in different parts of the world. These challenges often require extreme performances in terms of capacity, speed of delivery, energy consumption, environmental sustainability, as well as social and economic aspects.

A range of topics are covered, including: Navigation and ship operations, Ports and their operation, Port infrastructure, Safety and security, Emergency response, Pollution and the protection of the marine environment, Marine geodesy and cartography, Marine applications of geomatics and GIS, Satellite navigation, Routes planning and monitoring, Risk management, Naval technology, Freight handling, Multimode transport, Planning and management of ship traffic, Control and command systems, Human factors in navigation, Systems modelling and simulation, Management of coastal regions, Bulk shipping, Cruise shipping, Ro-ro and motorways of the sea, Meteorology and oceanography issues, Standardisation of maritime technology, Maritime education and training, Global socio-economic implications of maritime activities, Maritime policy and governance, International standards and regulations, Economic efforts of maritime transport, RAMS in maritime transport components.

ISBN: 978-1-78466-347-6 **eISBN: 978-1-78466-348-3**
Published 2019 / 284pp

www.ingramcontent.com/pod-product-compliance
Lightning Source LLC
LaVergne TN
LVHW062240200926
848484LV00006B/1582